THE FAMILY PROBLEM

New Internal Algebraic and Geometric Regularities

Gerald L. Fitzpatrick

Nova Scientific Press   *
Issaquah, Washington

A Monograph in Elementary-Particle Physics

Copyright ©1997 by Nova Scientific Press

Published by Nova Scientific Press
19226 S.E. 46th Place
Issaquah, WA 98027

All rights reserved. No part of this book may be reproduced, stored in a retrieval system, or transmitted, in any form or by any means, electronic, mechanical, photocopying, microfilming, recording, or otherwise, without written permission from the publisher. Individual readers of the material are, however, permitted to make fair use, including limited copying, for teaching or research purposes.

The publisher wishes to thank the following for permission to quote material: American Journal of Physics (vol. 51, no. 5, 1983, p. 398) for the (lower) epigraph on page 74 from *The Oersted Lecture* by J. A. Wheeler; American Institute of Physics and S. L. Glashow for the epigraph on page 50 from *The Charm of Physics* by S. L. Glashow, 1991, p. 96; The Balkan Agency for the epigraph on page 20 from *The Second Creation* by R. P. Crease and C. C. Mann, 1986, p. 187; Cambridge University Press (CUP) for the epigraphs on page 39 and 1 (upper) taken, respectively, from *Is The End in Sight for Theoretical Physics?* by S. W. Hawking, 1980, p. 23, and *Elementary Particles and the Laws of Physics* by R. P. Feynman and S. Weinberg, 1986, p. 10. We also thank CUP for the epigraphs on pages 1 (lower) and 74 (upper) taken, respectively, from p. 205 and p. 183, *Superstrings—A Theory of Everything?* edited by P. C. W. Davies and J. Brown, 1988. We thank W. H. Freeman and Company for the epigraph on page 67 from *Gravitation* by C. W. Misner, K. S. Thorne and J. A. Wheeler, 1973, p. 1215. We thank Addison-Wesley Longman, Inc. and Y. Ne'eman for the epigraph on page 34 taken from p. 304 of his book *Algebraic Theory of Particle Physics* (©1967 W. A. Benjamin Inc.). Finally, the publisher thanks M. L. Perl for permission to print the letter reproduced in the editor's foreword.

Library of Congress Catalog Card Number 96-92983
ISBN 0-9655695-0-0

Printed and bound by Maple-Vail Book Manufacturing Group, Binghamton, N.Y., USA.

The paper used in this publication meets the minimum requirements of American National Standards for Information Sciences—Permanence of Paper for Printed Library Materials, ANSI Z39.48-1984.

Editor's Foreword

Because the market for physics monographs is a relatively small one, it can be difficult, even for well-known authors, to find a publisher for this type of work. The original plan was to circumvent this problem by foregoing traditional hard-copy publication in favor of placing the manuscript in the Los Alamos National Laboratory e-Print Archive for High Energy Physics—Phenomenology. Only after the actual attempt to "put" the manuscript in the archive failed on April 2, 1996 was it learned that manuscripts must be accompanied by an "official report number from an academic institution." Subsequent requests by the author of this book (and others) to relax this policy have been unsuccessful.

> *I strongly recommend that you accept this paper by Gerald L. Fitzpatrick in the Los Alamos e-Print Archive. It certainly meets all your standards and I don't see why an institutional affiliation is necessary for such a paper.*
>
> *As a 1995 Nobel Laureate in Physics I am giving quite a number of talks this year. One of my subjects is the relation of theory to experiment. I will not be able to resist some mention of your slightly peculiar belief that being a member of an institution guarantees that a theoretical paper is worthwhile.*
>
> Martin L. Perl
> Stanford University
> April 23, 1996

Faced with these obstacles, and wishing to avoid further delays, it was finally decided late in October of 1996 to undertake self-publication. While this approach has many drawbacks, not least of which is the associated expense, it also has one very important advantage—the author/editor/publisher has total control of the publication process.

In this book, the author explores some of the algebraic, geometric and physical consequences of a new *organizing principle* for fundamental fermions (quarks and leptons). The essence of the new organizing principle is the idea that the standard-model concept of *scalar fermion-numbers* f can be generalized. In particular, a "generalized fermion-number," which consists of a 2 × 2 real *matrix* **F** that "acts" on an *internal 2-space*, instead of *space-time*, is taken to describe certain *internal* properties of fundamental fermions. This generalization automatically introduces internal degrees of freedom that "explain," among other things, family *replication* and the *number* of families.

Preface

$f \Rightarrow \mathbf{F}$

Parsimony: Economy in the use of means to an end; especially economy of explanation in conformity with Occam's razor.

Occam's razor: A scientific and philosophic rule that entities should not be multiplied unnecessarily, which is interpreted as requiring that the simplest of competing theories be preferred to the more complex or that explanations of unknown phenomena be sought first in terms of known quantities.

—Webster's Ninth New Collegiate Dictionary

I have written this book to bring together, in one place, a discussion of some of the algebraic, geometric and physical consequences of a new *organizing principle* for fundamental fermions (quarks and leptons). My starting point in elucidating this principle is the standard model of particle physics.

To address an important deficiency of the standard model, specifically the inability of this model to explain family "replication," I use the Cayley-Hamilton theorem for matrices to generalize the standard-model concept of scalar fermion-numbers f (i.e., $f_m = +1$ for fermions and $f_a = -1$ for antifermions). This theorem states that *every (square) matrix satisfies its characteristic equation*, which suggests that it may be possible to replace f, as expressed by an appropriate characteristic equation, by a more general matrix $\mathbf{F}$, provided f are the eigenvalues of $\mathbf{F}$. In particular, the essence of the new organizing principle, which "extends" the standard model, is this: assuming that the underlying physics (e.g., superstrings) permits such a "macroscopic" or "low"-energy description, *replace the scalar fermion-number f (a 1×1 real matrix) by a 2×2 real matrix $\mathbf{F}$, where f are the eigenvalues of $\mathbf{F}$, and $\mathbf{F}$ "acts" on some real, internal two-dimensional linear vector-space $V_2(\mathcal{R})$.* Flavor degrees of freedom (e.g., the "up"-"down"-type flavor dichotomy) and three families are found to be "automatic" consequences of this matrix generalization of f.

Using this simple generalization, I will show that the eigenvectors $\mathbf{Q}$ of $\mathbf{F}$ (i.e., $\mathbf{FQ} = f\mathbf{Q}$), together with certain pairs of linearly-independent vectors ($\mathbf{U}$ and $\mathbf{V}$) that resolve $\mathbf{Q}$ (i.e., $\mathbf{Q} = \mathbf{U} + \mathbf{V}$), namely, various non-Euclidean "vector-triads" ($\mathbf{Q}, \mathbf{U}, \mathbf{V}$)—these are the analogs of Euclidean triangles—serve to represent both individual *flavors* and *flavor-doublets* of fundamental fermions. Moreover, it will be shown that *different* resolutions of $\mathbf{Q}$ represent *different* flavor doublets having the *same* $\mathbf{Q}$. Thus, I am led to "predict" (*ex post facto*) that flavor doublets of fundamental fermions having the same $\mathbf{Q}$ are *replicated*. I will also show that there are only *two* types of $\mathbf{Q}$-vectors—one for *quarks* and one for *leptons*—where the two components of $\mathbf{Q}$ are the two *electric* charges of flavor-doublet members. Finally, I will show that there are only *three* physically acceptable ways these $\mathbf{Q}$-vectors can be resolved for quarks (leptons), meaning that there are only *three* flavor-doublets of quarks (leptons) and *three* quark-lepton families.

While the new organizing principle is simple, and while it is apparently *not* in conflict with quantum mechanics or relativity, it is also unconventional, i.e., $\mathbf{F}$ is *not* a spacetime-dependent quantum *operator* and its eigenvectors are *not* states in a Hilbert space. In particular, $\mathbf{F}$ (like f) is a generally non-Hermitian real matrix whose components are required to be global charge-conjugation-reversing ($\mathbf{C}$-reversing), or charge-like, quantum observables. Moreover, the 2-space on which $\mathbf{F}$ "acts" has some unusual properties.

Because the 2-space is supposed to be an *internal* space it is necessary to assume that it is associated with, or "carried" (in spacetime) by, *every* individual fundamental-fermion. In this sense, the 2-space is analogous to an isospin 3-space and the 2-vectors in this space, like the total isospin 3-vector $\mathbf{T}$, represent certain *intrinsic* properties of individual fundamental-fermions (and their associated flavor doublets). But, there are important differences between these 2-vectors and 3-vectors such as $\mathbf{T}$.

In the first place, a self-consistent description of antimatter requires the 2-space to be "Lorentzian," i.e., *non-Euclidean*, whereas isospin space is a *Euclidean* 3-space. Moreover, no mathematical (or physical) meaning can be assigned to the *matrix* $\mathbf{F}$, its *characteristic equation*, or its *eigenvectors* $\mathbf{Q}$ and *eigenvalues* f, unless: the scalar components of the associated vectors and matrices can all be known, in principle, precisely and *simultaneously*. In particular, these components must be mutually-commuting (compatible) quantum observables. By contrast, the measurement of one component of $\mathbf{T}$ always interferes with the measurement of any other component of $\mathbf{T}$ at a later time (i.e., the components of $\mathbf{T}$ are incompatible and do not commute).

I find it necessary to require that all of the aforementioned 2-scalars are mutually-commuting, $\mathbf{C}$-reversing *internal* "charges" similar to the diagonal charges of a Cartan subalgebra [they are also $U(1)$-type charges that generate phase transformations]. But, while a non-Euclidean scalar-product or C-reversing "charge" such as $\mathbf{Q} \bullet \mathbf{Q} = \mathbf{Q}^2$ (not a probability density!) is *additive* for a "composite" of two different fundamental fermions (each carrying $\mathbf{Q}$), the corresponding $\mathbf{Q}$-vectors are not. For example, if these vectors were additive, the composite in question would be required to carry a nonsensical "charge" $(\mathbf{Q}+\mathbf{Q})^2 = 4\mathbf{Q}^2$, in adition to the expected charge $\mathbf{Q}^2 + \mathbf{Q}^2 = 2\mathbf{Q}^2$. Hence, 2-vectors of a given type (i.e.; $\mathbf{Q}, \mathbf{U}$ or $\mathbf{V}$) for two *different* particles are *not* permitted to combine or "interfere"—only their corresponding 2-scalar charges can be combined. From an algebraic standpoint, this means that the 2-*spaces associated with (carried by) two different fundamental fermions are effectively isolated from one another*. And, this situation is roughly the converse of that for a vector such as $\mathbf{T}$, where $\mathbf{T}^2$ is neither additive nor a charge-like scalar (i.e., it is C-invariant), but $\mathbf{T}$ is "additive" for two different particles, i.e., $\mathbf{T}_1 + \mathbf{T}_2 = \mathbf{T}$ and $\mathbf{T}^2 = \mathbf{T}_1^2 + 2\mathbf{T}_1 \bullet \mathbf{T}_2 + \mathbf{T}_2^2$.

Just as the "diagonal" charges of a Cartan subalgebra can be used to define *simultaneous eigenstates*, certain appropriate sets of these global mutually-commuting charges (i.e., flavor-defining charge-like quantum numbers) will be used to define *simultaneous flavor-eigenstates*. It will be argued that, in the presence of strong, electromagnetic and weak neutral-current interactions (e.g., pair production), these flavor-defining quantum numbers are *exactly* conserved, i.e., they are "good" quantum numbers. Only in the case of weak charge-changing interactions (and in gravitational collapse) will *some* (or *most*) of these charges be destroyed. In this way, a new global *internal* description of *flavors*, *flavor-doublets*, and *families* is achieved.

While it is possible, in principle, to start with the new organizing principle and *derive* the first family of fundamental fermions, in practice the physical interpretation of geometric objects in the 2-space is "forced" when this approach is taken. Instead, it is easier and more natural to assume the existence of *first-family* fermions (i.e., their colors, flavors and global quantum numbers), to show that they can be described using the new organizing principle and then, armed with this success, proceed to derive the *second* and *third* families from the organizing principle.

In Chapter 1, a general discussion of the mathematical and physical consequences of "replacing" spacetime-dependent scalar fermion-numbers by 2×2 real matrices is presented. The framework for describing flavors, flavor-doublets and families is established here. In Chapter 2, first-family data are used to provide a partial justification for introducing the new internal 2-space, and to describe, in detail, how this space can be used to represent flavors, flavor-doublets and families. For example, it is shown that *baryon numbers* and *lepton numbers* can be represented as 2-scalars in this space. In Chapter 3, the new internal-space is depicted graphically on a Euclidean plane (E2). The four first-family "up"- and "down"-type flavor, or mass, eigenstates (flavors) serve to define four *preferred* "flavor-directions" in the 2-space. These are *two* oppositely-directed flavor-directions for each of *two* coordinate axes. In Chapter 4, it is demonstrated that algebraic and geometric properties of the new internal-space, together with certain standard-model constraints, lead to an effective "quantization" of various global flavor-defining scalars (i.e., the scalar components of the vectors **U** and **V**). This "quantization" leads, in turn, to the *ex post facto* prediction that there are a total of only *twelve* distinguishable flavors and *three* distinguishable families of fundamental fermions. In Chapter 5, the exact mathematical form of the matrix **F** is deduced and the *electric* charges of fundamental fermions are shown to be the components of the eigenvectors of **F**. In Chapter 6, it is demonstrated that the new formalism has a natural place for a hierarchical relationship between three families. For example, a matrix **R** exists that generates three pairs of **U**-type vectors in the hierarchy: $\pm\mathbf{U}$, $\pm\mathbf{RU}$ and $\pm\mathbf{R}^2\mathbf{U}$. Here, the pair of vectors $\pm\mathbf{U}$ describe the first family, $\pm\mathbf{RU}$ describe the second family and $\pm\mathbf{R}^2\mathbf{U}$ describe the third family, where plus signs describe quark flavor-doublets, while minus signs describe the corresponding "inverted" lepton flavor-doublets. In Chapter 7, the global flavor-defining charge labels and flavor, or mass, eigenstates allowed by the new description are tabulated. In Chapter 8, I speculate on the reasons for the apparent success of the Cayley-Hamilton generalization $f \Rightarrow \mathbf{F}$ and summarize "predictions."

I hope that this book will serve to engender further (similar) research and that many questions left unanswered in the present work will, thereby, be addressed and ultimately answered.

I am indebted to L. P. Horwitz and J. A. Wheeler for encouragement and for many helpful comments and suggestions during the early stages of this work. I thank R. F. Holub, P. K. Smrz, H. P. Noyes, R. D. Peccei, D. K. Thome and T. R. Morgan for discussions. Finally, I thank R. L. Skaugset for formalizing book design-drawings and M. E. Sheetz for very intelligent and skillful conversion of rough manuscript to final product.

Issaquah, Washington *Gerald Fitzpatrick*
December 1996

Contents

    Preface   v

1 Introduction and Mathematical Preliminaries   1
   1.1 Fermion Numbers in the Standard Model 2
   1.2 A New Organizing Principle for Fundamental Fermions 3
       1.2.1 A Cayley-Hamilton Based Generalization 3
       1.2.2 Preliminary Statement of the New Organizing Principle   6
       1.2.3 The Form of the Matrix **F** 7
       1.2.4 **F** as a Special Case of the Matrix τ 9
       1.2.5 The Metric and Scalar-Products 11
       1.2.6 Restatement of the New Organizing Principle 11
   1.3 Preliminary Physical Interpretation 13
       1.3.1 Mutually-Commuting Global Charges and the 2-Space   14
   1.4 Representing Flavor Eigenstates and Flavor Doublets 16
       1.4.1 Flavor Doublets 17
       1.4.2 Flavor Conservation 18
       1.4.3 Effective "Quantization" 19

2 The 2-Space Description of First-Family Fermions   20
   2.1 Family Definition and the Family Hierarchy 20
   2.2 Labeling and Subscript Conventions 21
   2.3 B and L as Functions of Electric Charge Q 22
       2.3.1 B and L as Linear Functions of Electric Charge Q . . . 24
       2.3.2 General B and L Values as Quadratic Functions of Q . 24
   2.4 Scalar-Products of 2-Vectors 25
       2.4.1 General B and L Values Expressed as Scalar-Products   26
       2.4.2 Scalar Fermion-Numbers f and the Matrix **F** 27
       2.4.3 Orthogonality of Matter and Antimatter 28

2.5		A Preliminary Geometric Interpretation of Three Quantum Dichotomies	29
	2.5.1	The "up"-"down" Flavor Dichotomy	29
	2.5.2	The Matter-Antimatter Dichotomy	29
	2.5.3	The Quark-Lepton "Dichotomy"	30
2.6		Other Quantum Numbers and Families in the 2-Space	31
	2.6.1	Global Flavor-Defining Charge Labels	31
	2.6.2	Flavor Eigenstates	32
	2.6.3	Effective "Quantization"	33

3 Graphical Representation of the New 2-Space — 34
- 3.1 Flavor and Antiflavor Directions 34
- 3.2 Unphysical Regions of the 2-Space 35
- 3.3 Other Geometric and Topological Properties of the 2-Space . . 37

4 Effective "Quantization" of the U-Vector Components and Three Families — 39
- 4.1 The Vectors **U** and **V** 39
  - 4.1.1 The Form of **U** and **V** 40
- 4.2 The Six Physically Acceptable **U**-Vectors 42
  - 4.2.1 The First Family 42
  - 4.2.2 The Second and Third Families 44
  - 4.2.3 Strangeness and Truth 46
  - 4.2.4 Family Kinship—Preliminary Considerations 48

5 Effective "Quantization" of the Q_q and Q_ℓ Vector Components — 50
- 5.1 New Labeling and Subscript Conventions 50
- 5.2 Effective "Quantization" of v and v' 52
  - 5.2.1 Strong Interactions and the Parameters v and v'. . . . 53
- 5.3 An Interpretation of the Components of **F** 56
- 5.4 Symmetries and Asymmetries Associated With the Matrix **F** 56
  - 5.4.1 Symmetry Under **F** or Symmetry Class I 57
  - 5.4.2 Asymmetry Under **F** or Symmetry Class II 57
  - 5.4.3 Symmetry "Evolution" 57

CONTENTS

6   A Hierarchy Among Families Revealed   **59**
    6.1   Representation of Flavor Doublets by **U**-Vectors 59
    6.2   The Family Hierarchy and the Matrix **R** 61
    6.3   An Interpretation of the Components of **R** 63
    6.4   The Four Main Kinds of 2-Space Transformations 66

7   Flavor Eigenstates and Their Mutually-Commuting Charge Labels   **67**
    7.1   The new Global Charge Relations 71
    7.2   Global Conservation Laws 72

8   Why Does the Generalization $f \Rightarrow \mathbf{F}$ Describe Fundamental Fermions?   **74**
    8.1   Particle Physics at the Planck Level. 76
    8.2   Planck-Level Physics and **F**. 77

Appendix A. Three Quantum "Dichotomies" and $SU(5)$.   **80**

Appendix B. The Matter-Antimatter "Reflection".   **83**

Appendix C. Identifying the Symmetry Class of Flavor Doublets.   **85**

Footnotes and References.   **88**

Index.   **95**

List of Tables

Table I. Labeling and Subscript Conventions for First-Family Fermions and Antifermions. 23

Table II. Summary of Global Transformations of Flavor Doublets. . 66

Table III. Quadrant IV Global Geometric Quantities and Quarks. . 68

Table IV. Quadrant II Global Geometric Quantities and Leptons. . 69

Table V. The Mutually-Commuting Flavor-Defining Global Charge Labels or Quantum Numbers and Flavor, or Mass, Eigenstates (ket Vectors) for Quarks. 70

Table VI. The Mutually-Commuting Flavor-Defining Global Charge Labels or Quantum Numbers and Flavor, or Mass, Eigenstates (ket Vectors) for Leptons. 70

Table VII. Quantized **C**-Reversing 2-Scalars. 72

List of Figures

Figure 3.1. An abstract, internal non-Euclidean 2-space 36

Figure 3.2. Additional geometrical and topological aspects of the 2-space 38

Figure 4.1. The six physically acceptable **U**-vectors 43

Figure 4.2. Anticipating... **U**-vectors, their components (u-charge) and the scalar-product $\mathbf{U}^2$. 47

So the requirements of positive energies and relativity force us to allow creation and annihilation of pairs of particles, one of which travels backwards in time.

Yes, indeed, there are wonderful principles, ideas of valence, sound, pressure, and many other organizing principles, which help to understand a complex situation.

—R. P. Feynman

1. Introduction and Mathematical Preliminaries

Both the cosmological evidence[1] on helium nucleosynthesis in the "big bang," and the measured Z^0 decay width[2] in particle-physics experiments, are consistent with a maximum of $N_\nu = 3$ low (or zero) mass neutrino flavors, namely, the ν_e, ν_μ, and ν_τ neutrinos. These observations strongly imply that only three "conventional" quark-lepton families exist in nature.

The recent discovery of the top quark[3] appears to complete the list of "fundamental" fermions (quarks and leptons) by completing the third family. Presumably all "ground state" fundamental-fermions, with the possible exception of certain hypothetical superpartners to existing fundamental-bosons (e.g., the photino), have now been observed either "directly" or indirectly.

The aforementioned experimental successes are tempered by the fact that the standard model of particle physics,[4] based on $SU(3)_c \times SU(2)_L \times U(1)_Y$, provides only a partial explanation for the existence of individual families of fundamental fermions and completely fails to explain their observed threefold "*replication.*" Moreover, extensions of the standard-model—these include grand unified theories[4-9] (GUTs) such as $SU(5)$, family gauge-symmetries,[10] composite models,[11,12] models that involve random dynamics at the Planck level[13] and string theories,[14] to mention a few—either fail to provide an explanation, or the explanations they do provide raise as many questions as they answer. Finally, while the experimental evidence certainly rules out additional families associated with *light* neutrinos, it does not rule out families associated with very *massive* neutrinos. Are such families likely to exist or are there reasons why they should not?

The purpose of this book is to propose a new *organizing principle* for fundamental fermions, i.e., an "extension" of the standard model based on a mathematical generalization of the standard-model concept of conserved scalar *fermion-numbers*. This generalization will lead to new *internal* degrees of freedom (i.e., new internal, global quantum-numbers) and a new description of families that is more or less complementary to the standard-model description but, unlike the standard-model description, provides a "simple" explanation for family replication and the number of families. To see how fermion numbers might be generalized, and how this might come about, first let us review how they are sometimes described in the standard model.

1.1 Fermion Numbers in the Standard Model

Because standard quantum field-theories require the Lagrangian describing interacting fermions to be both phase (gauge) and Lorentz invariant, it can be shown that the total fermion-number must be conserved.[15-19] That is, *the number of fermions minus the number of antifermions involved in an interaction must be a constant of the motion.* Hence, both the fermion number f and the electric charge q of an "isolated" fermion or antifermion, are "good" charge-like quantum numbers that can be used to provide a partial representation of the corresponding quantum states (Ref. 15, p. 12). But, *the "charge" f, unlike the electric charge q, or local color-gauge-charges, is strictly global in nature, i.e., insofar as we know it is not coupled to any field of force.*

Because the total fermion-number is independent of time and space, it can be evaluated far from the interaction region where the physical situation is well described by single-particle fermion and antifermion quantum states in Hilbert space, namely, $|\mathbf{p}\rangle_s$ and $|\bar{\mathbf{p}}\rangle_s$, respectively. In particular, the observable fermion-numbers of individual fermions or antifermions of *any* flavor may be represented by a mathematically well-defined, but physically unobservable, *quantum operator* $\mathbf{F}(\text{op})$. This operator, which is constructed from an appropriate combination of creation and annihilation operators, acts on mathematically well-defined, but physically unobservable, single-particle quantum states to produce their associated fermion-numbers (see Ref. 15, pp. 23–25) thus:

$$\mathbf{F}(\text{op})|\mathbf{p}\rangle_s = f_m |\mathbf{p}\rangle_s \tag{1.1}$$

or

$$\mathbf{F}(\text{op})|\bar{\mathbf{p}}\rangle_s = f_a |\bar{\mathbf{p}}\rangle_s. \tag{1.2}$$

By definition, $f_m = +1$ is the fermion number for a single "isolated" *fermion*, and $f_a = -1$ is the fermion number for the corresponding *antifermion*.

1.2. A NEW ORGANIZING PRINCIPLE

In the next section, equations for f_m and f_a similar in appearance to (1.1) and (1.2), but very different in mathematical and physical content, will be introduced. In addition to f_m and f_a, these equations involve *internal* global "observables" and a corresponding set of "good" (conserved) quantum numbers that will be used to represent fundamental-fermion flavor-eigenstates, flavor-doublets and families.

1.2 A New Organizing Principle for Fundamental Fermions

While the foregoing description of scalar fermion-numbers serves to distinguish fermions and antifermions in a *flat* spacetime setting, fermion numbers by themselves do *not* distinguish between quarks and leptons, between "up"- and "down"-type flavors or between flavors in different families. Additional *internal* physical properties of fundamental fermions (e.g., electric charge, color and flavor) are required to break these "symmetries" and describe experimental observations. Some of these properties are provided by the standard model and some, such as the (global) flavor-defining quantum numbers for family members beyond the first family, are not. This circumstance leads one to wonder if there is some *internal* (as opposed to spacetime) mathematical generalization of scalar fermion-numbers f (i.e., f_m or f_a) that not only distinguishes fermions and antifermions, but also distinguishes different flavors and families.

It will be argued (see Sec. 8.0) that such an *internal* picture could arise naturally in going from a description of matter in a *flat* gravity-free spacetime, to a description of matter in a *curved* spacetime (e.g., quantum gravity). In any case, to be successful, such a phenomenological generalization would have to be more or less compatible with the standard model, but at the same time it would have to provide an explanation for family replication. An important theorem from the algebra of matrices provides a very simple, and natural, mathematical generalization of f that seems to have many of the desired characteristics.

1.2.1 A Cayley-Hamilton Based Generalization

The Cayley-Hamilton theorem (see Ref. 20, p. 136) states that *every (square) matrix satisfies its characteristic equation.* From a mathematical perspective, this theorem suggests that if the underlying physics (i.e., *internal* properties of fundamental fermions) permits the scalar fermion-numbers f (1×1 matrices) to be described as eigenvalues of some square matrix **F** [where $\mathbf{F} \neq \mathbf{F}(\text{op})$, and the components of **F**, like f, are conserved fundamental-fermion *observables*], then f can be replaced in the characteristic equation of **F**, by a more general matrix **F**. In particular, suppose that the scalar fermion-numbers f (f_m and f_a) defined in a *spacetime* setting by Eqs. (1.1) and (1.2) are, at the same time, eigenvalues of some *real* 2×2 square matrix **F**, which acts on an *internal* two-dimensional linear vector-space $\mathbf{V}_2(\mathcal{F})$ defined on the real-number field $\mathcal{F} = \mathcal{R}$.[21] Then, the characteristic equation of **F** is, by definition,

$$det\,(\mathbf{F} - f\,\mathbf{I}_2) = 0, \tag{1.3}$$

where $\mathbf{I}_2$ is the 2×2 identity matrix. This equation ensures that the associated system of homogeneous linear equations

$$(\mathbf{F} - f\,\mathbf{I}_2)\mathbf{Q} = 0, \tag{1.4}$$

can be solved (nontrivially) for the eigenvectors of **F**, namely, the 2-component column vectors or 2-vectors **Q**, where $\mathbf{Q} \neq |\mathbf{p}\rangle_s$ or $|\bar{\mathbf{p}}\rangle_s$, and the scalar components of **Q**, like those of **F**, are taken to be conserved, global *observables* associated in some way with fundamental fermions.

Since Eqs. (1.3) and (1.4) are supposed to describe only *matter* or *antimatter*, the maximum number of linearly-independent eigenvectors **Q** of **F** (corresponding to f) must be equal to the number of distinct eigenvalues of **F**, which is two. These requirements will be satisfied provided **F** is similar to a *diagonal* matrix $\mathbf{F}_{\text{diag}}$ (see Ref. 20, p. 173). That is, there must exist a nonsingular matrix **P** and a *similarity transformation* $\mathbf{PFP}^{-1} = \mathbf{F}_{\text{diag}}$ (see Ref. 20, p. 143). Since the eigenvalues of **F** were previously *defined* to be f_m and f_a these numbers, necessarily, form the diagonal elements of $\mathbf{F}_{\text{diag}}$, and the off-diagonal elements of $\mathbf{F}_{\text{diag}}$ are, by definition, zero (Ref. 20, p. 169). Two choices for $\mathbf{F}_{\text{diag}}$ are possible depending on the relative positions of f_m and f_a on the diagonal. Throughout this book I employ the following specific form for $\mathbf{F}_{\text{diag}}$:

$$\mathbf{F}_{\text{diag}} = \begin{pmatrix} f_m & 0 \\ 0 & f_a \end{pmatrix}. \tag{1.5}$$

The expansion of (1.3) yields the characteristic equation of **F**, namely,

$$f^2 - f\,tr\,\mathbf{F} + det\,\mathbf{F} = 0, \tag{1.6}$$

1.2. A NEW ORGANIZING PRINCIPLE

which simplifies to $f^2 = 1$ since the invariant trace and determinant of $\mathbf{F}$ are given, respectively, by (note that $tr\,\mathbf{F} = tr\,\mathbf{F}_{\text{diag}}$, $det\,\mathbf{F} = det\,\mathbf{F}_{\text{diag}}$)

$$tr\,\mathbf{F} = f_m + f_a = 0 \tag{1.7}$$

and

$$det\,\mathbf{F} = f_m f_a = -1. \tag{1.8}$$

Then, since $f^2 = 1$ is the characteristic equation of $\mathbf{F}$, and since $det\,\mathbf{F}$ is nonzero, $\mathbf{F}$ must be some nonsingular 2×2 discrete matrix that, according to the Cayley-Hamilton theorem, satisfies the *same* equation, namely,

$$\mathbf{F}^2 = \mathbf{I}_2. \tag{1.9}$$

In other words, *both f and* $\mathbf{F}$ *are square-roots of unity*. This means that $\mathbf{F}$ is equal to its multiplicative inverse $\mathbf{F}^{-1}$, which is necessarily *unique* because $\mathbf{F}$ is *nonsingular* (see Ref. 20, p. 41).

Another way to say these things is to say that f_m and f_a are *generators* of one-dimensional representations of the simplest possible of nontrivial abstract groups, namely, the finite two-element discrete group of involutions called Z_2. In particular, the two irreducible one-dimensional representations (Ref. 22, p. 105) are $Z_2(f) = \{f, 1\}$ or $Z_2(1) = \{1, 1\}$ and $Z_2(-1) = \{-1, 1\}$. Similarly, the matrices $\mathbf{F}$ will be the *generators* of some higher-dimensional reducible representation of the group Z_2 (see Ref. 21 and Ref. 22, p. 105), namely,

$$Z_2(\mathbf{F}) = \{\mathbf{F}, \mathbf{I}_2\}. \tag{1.10}$$

In the foregoing, the matrix $\mathbf{F}$ was arbitrarily assumed to be *real*. But, it can be shown that the most general 2×2 Hermitian matrix, namely,[23]

$$\tau = \begin{pmatrix} a & \sqrt{1-a^2}\,e^{i\phi} \\ \sqrt{1-a^2}\,e^{-i\phi} & -a \end{pmatrix}, \tag{1.11}$$

where a is a real number and $a^2 \leq 1$, also satisfies Eqs. (1.7), (1.8) and (1.9). Moreover, it is easily demonstrated that the Hermitian matrices known as Pauli matrices, which are generators of $SU(2)$, are special cases of (1.11). It will also be shown later that the real (generally non-Hermitian) matrices $\mathbf{F}$ can be obtained from (1.11) provided a is real and $a^2 \geq 1$.

1.2.2 Preliminary Statement of the New Organizing Principle

Let us recapitulate the main proposal in Sec. 1.2.1, and also determine some of its inevitable algebraic, geometric and physical consequences. It is being proposed that the matter-antimatter dichotomy of fundamental fermions can be given both a conventional spacetime-dependent quantum description as in Eqs. (1.1) and (1.2), and/or it can be given this new *internal* 2-space description as in Eqs. (1.3) and (1.4). In particular, any acceptable 2×2 matrix $\mathbf{F}$ possesses just *two*, *real* linearly-independent eigenvectors, call them $\mathbf{Q}$ and $\mathbf{Q}^c$, corresponding to the two *real* eigenvalues f_m and f_a, respectively. Therefore, the matrix $\mathbf{F}$ can be thought of as "producing" the conventional single-particle fermion numbers f_m and f_a via the 2-space eigenvalue equations or transformation "laws"

$$\mathbf{FQ} = f_m \mathbf{Q} \tag{1.12}$$

and

$$\mathbf{FQ}^c = f_a \mathbf{Q}^c, \tag{1.13}$$

respectively.

The 2-vector $\mathbf{Q}$ (and its scalar components) describes some as yet undetermined aspect of *matter*, while the linearly-independent 2-vector $\mathbf{Q}^c$ describes its antimatter counterpart. The superscript c on $\mathbf{Q}^c$ is merely a label signifying antimatter. It is not an exponent or a symbol for complex conjugation. As such, it signifies only that *the 2-vectors $\mathbf{Q}$ and $\mathbf{Q}^c$ are real vectors associated with ("carried by," "representing," etc.) individual fundamental-fermions or antifermions, respectively*, not state vectors in some Hilbert space.

The two-dimensional character of this space automatically provides a natural way to introduce "degrees of freedom" beyond *matter* and *antimatter* into the description of fundamental fermions. It will be shown that this 2-space naturally describes "up"- and "down"-type flavor degrees of freedom, namely, the "up"-"down"-type flavor dichotomy of fundamental fermions (i.e., fundamental fermions come in flavor doublets). In particular, *the same 2-vectors $\mathbf{Q}$ and $\mathbf{Q}^c$ will also be associated with (represent) flavor doublets and their antiparticle counterparts, respectively*. Finally, it will be shown that the quark-lepton "dichotomy" [the dichotomy of strongly-interacting versus electroweakly-interacting (only) fundamental fermions] is associated, in part, with the inversion $\mathbf{I} = -\mathbf{I}_2$.

The proposed organizing principle for fundamental fermions takes the following preliminary form:

1.2. A NEW ORGANIZING PRINCIPLE

Individual fundamental-fermions, and their associated flavor-doublets, can be described by "generalized fermion-numbers" consisting of 2×2 real matrices $\mathbf{F}$. These transformation matrices "act" on an internal, real two-dimensional vector-space through an eigenvalue equation $\mathbf{FQ} = f_m \mathbf{Q}$ or $\mathbf{FQ}^c = f_a \mathbf{Q}^c$, where the scalar fermion-number $f_m = +1$ describes matter ($\mathbf{Q}$) and $f_a = -1$ describes antimatter ($\mathbf{Q}^c$). Specifically, $\mathbf{Q}(\mathbf{Q}^c)$ is a real 2-vector that represents both individual fundamental-fermions (antifermions) and their associated "up"-"down"-type flavor-doublets (antiflavor-doublets).

The Cayley-Hamilton "replacement" $f \Rightarrow \mathbf{F}$, which is the essence of the new organizing principle, does *not* mean that f_m or f_a are to be replaced by $\mathbf{F}$ wherever these quantum numbers happen to appear in an equation of physics (e.g., $\mathbf{F}$ does *not* replace f_m or f_a in the operator equations of Sec. 1.1). Instead, it means that $\mathbf{F}$ is to be used in some entirely new description of fundamental fermions where $\mathbf{F}$ "acts" on the associated internal 2-space in which other matrices, vectors, and scalars [i.e., "good" (conserved) flavor-defining quantum numbers] also play a role in the description.

After we have determined exactly how to distinguish quarks from leptons in this 2-space, and also determined the metric, we will restate the new organizing principle in its final form. At that time, we will also indicate how this principle can be used to "explain" family replication and to "predict" (*ex post facto*) the number of families.

1.2.3 The Form of the Matrix F

Restricting $\mathbf{F}$ to be a square-root of unity as in (1.9) does *not* determine $\mathbf{F}$. To do this we must make certain additional assumptions. Given that the eigenvalues of $\mathbf{F}$ for an "isolated" fermion are *conserved* charge-conjugation-reversing (C-reversing) quantum observables (i.e., the "good" charge-like quantum numbers $f_m = -f_a = 1$), *the scalar components of* $\mathbf{F}$, $\mathbf{Q}$ *and* $\mathbf{Q}^c$ *are likewise taken to be conserved* C-*reversing quantum observables*. In particular, I assume that these scalars are C-reversing charge-like quantum numbers or "charges" in both *spacetime* (i.e., they are Lorentz 4-scalars) and the *internal* 2-space (i.e., they are 2-scalars). These, and other conserved flavor-defining global "charges," will be used later to distinguish fundamental-fermions and *define* simultaneous flavor-eigenstates.

Because the matrix $\mathbf{F}$ is C-reversing in the sense described immediately above, a 2×2 *real* nonsingular matrix $\mathbf{X}$, together with its multiplicative inverse $\mathbf{X}^{-1}$, should exist such that $\mathbf{F}$ is transformed in the 2-space to its C-reversed counterpart $\mathbf{F}^c$ by the similarity transformation (Ref. 20, p. 143)

$$\mathbf{X}\,\mathbf{F}\,\mathbf{X}^{-1} = \mathbf{F}^c = -\mathbf{F}. \tag{1.14}$$

Because $\mathbf{X}$ is supposed to be analogous to charge conjugation $\mathbf{C}$, two applications of $\mathbf{X}$ should return a *real* 2-vector to its original state.[24] Therefore, $\mathbf{X} = \mathbf{X}^{-1}$ or

$$\mathbf{X}^2 = \mathbf{I}_2, \tag{1.15}$$

is required. Hence, $\mathbf{X}$ and $\mathbf{F}$ both generate the group Z_2.

To find the unknown matrix $\mathbf{X}$, first diagonalize the matrix $\mathbf{F}$ and then substitute $\mathbf{F}_{\text{diag}}$ in (1.14). Using Eqs. (1.14), (1.15) and (1.5) it can be shown that

$$\mathbf{X} = \begin{pmatrix} 0 & 1/d \\ d & 0 \end{pmatrix}, \tag{1.16}$$

where d is some real number. Since the matrix $\mathbf{X}$ is supposed to transform $\mathbf{F}$ to its C-reversed counterpart as in (1.14), a self-consistent physical picture requires that this same matrix $\mathbf{X}$ must transform the column-vectors $\mathbf{Q}$ and $\mathbf{Q}^c$ to their C-reversed counterparts, namely,

$$\mathbf{X}\,\mathbf{Q} = \mathbf{Q}^c$$

and

$$\mathbf{X}\,\mathbf{Q}^c = \mathbf{Q},$$

respectively. Given that the components of $\mathbf{Q}$ and $\mathbf{Q}^c$ are C-reversing "charges" the signs, but not the magnitudes, of these components must change under $\mathbf{X}$. Hence, one finds that all constraints on $\mathbf{X}$ are met if $d = -1$ or

$$\mathbf{X} = -\boldsymbol{\sigma}_x = \begin{pmatrix} 0 & -1 \\ -1 & 0 \end{pmatrix}, \tag{1.17}$$

where $\boldsymbol{\sigma}_x$ is one of the familiar Pauli-matrices. Thus, if

$$\mathbf{Q} = \{q_1, q_2\} \tag{1.18}$$

1.2. A NEW ORGANIZING PRINCIPLE

is a column vector, and q_1, q_2 are its component "charges," then

$$\mathbf{Q}^c = \{-q_2, -q_1\}. \tag{1.19}$$

This last result demonstrates that $\mathbf{X}$ not only changes the *signs*, but also reverses the *order* of column-vector component charges. Hence, $\mathbf{Q} \neq -\mathbf{Q}^c$.

In general, $\mathbf{F}$ is *not* a diagonal matrix, and since the similarity transformation defined by Eqs. (1.14) and (1.17) is equivalent to an exchange of rows followed by an exchange of columns, or vice versa, $\mathbf{F}$ must have the general form:

$$\mathbf{F} = \begin{pmatrix} a & b \\ -b & -a \end{pmatrix}. \tag{1.20}$$

But, since $det\,\mathbf{F} = f_m f_a = -1$ from (1.8), $a^2 - b^2 = 1$ for all $\mathbf{F}$. Making the substitutions $a = \cosh v$ and $b = \pm \sinh v$ in (1.20) leads to the following general expression for $\mathbf{F}$, namely,

$$\mathbf{F} = \mathbf{F}(v) = \begin{pmatrix} \cosh v & \pm \sinh v \\ \mp \sinh v & -\cosh v \end{pmatrix}, \tag{1.21}$$

where the boundary condition (1.5) is clearly satisfied since $\mathbf{F}(0) = \mathbf{F}_{\text{diag}}$. We will show later that $\mathbf{F}(v)$ has physical significance only when the real, dimensionless-parameter $v \geq 0$, and when the upper signs in (1.21) are chosen.

It is apparent that different choices for the dimensionless-parameter v give different matrices $\mathbf{F}(v)$. We will show that there are only *two* possible choices for v—hence that there are only *four* distinct, fixed (constant) eigenvectors $\mathbf{Q}_q$, $\mathbf{Q}_q^c$, $\mathbf{Q}_\ell$ and $\mathbf{Q}_\ell^c$ associated with $\mathbf{F}(v)$. This circumstance, together with the 2-space *inversion*, provides a natural basis for distinguishing *quarks* (q) and *leptons* (ℓ). Thus, the new organizing principle, which already incorporates the matter-antimatter and "up"-"down"-type flavor dichotomies, also provides a natural way to incorporate the quark-lepton "dichotomy."

1.2.4 F as a Special Case of the Matrix τ

By making the substitutions $a = \cosh v$ and $e^{i\phi} = -i$ in (1.11), the Hermitian matrix τ becomes the real, generally non-Hermitian (when $v \neq 0$) matrix $\mathbf{F}$ of equation (1.21). Clearly, the matrix $\mathbf{F}$ and the familiar Pauli-matrices are closely related. But, a Hermitian matrix "operator" such as τ acts on a Hilbert space, whereas the real, generally non-Hermitian matrix **F** "acts" on a real, internal, 2-space. Thus, the two *physical* situations described by these matrices are entirely different in spite of their *mathematical* kinship.

It is clear that the transformation $\mathbf{F}(v)$ in (1.21) can be written as a product of two factors, namely, a simple *discrete* transformation $\boldsymbol{\sigma}_z$ and a *continuous* transformation $\mathbf{T}(v)$. That is, when the upper signs in (1.21) are chosen,

$$\mathbf{F}(v) = \boldsymbol{\sigma}_z \mathbf{T}(v), \tag{1.22}$$

where $\boldsymbol{\sigma}_z$ is the 2×2 Pauli matrix

$$\boldsymbol{\sigma}_z = \begin{pmatrix} 1 & 0 \\ 0 & -1 \end{pmatrix}, \tag{1.23}$$

and $\mathbf{T}(v)$ is a 2×2 matrix (Ref. 22, p. 128) with unit-determinant representing an element of the so-called continuous pseudorotation or "Lorentz" group $SO(1,1)$, namely,

$$\mathbf{T}(v) = \begin{pmatrix} \cosh v & \sinh v \\ \sinh v & \cosh v \end{pmatrix}. \tag{1.24}$$

When the lower signs in (1.21) are chosen,

$$\mathbf{F}(v) = \mathbf{T}(v)\boldsymbol{\sigma}_z. \tag{1.25}$$

Finally, because $\mathbf{T}(v)$ and $\boldsymbol{\sigma}_z$ *both* preserve "Lorentzian" quadratic forms (see below), the 2-space metric **g** is, necessarily, "Lorentzian" or non-Euclidean:

$$\mathbf{g} = \begin{pmatrix} 1 & 0 \\ 0 & -1 \end{pmatrix}. \tag{1.26}$$

The fact that the metric is "Lorentzian" means that whenever *any* $\mathbf{F}(v)$-matrix acts on a column vector $\{\alpha, \beta\}$, the associated "Lorentzian" quadratic form $\alpha^2 - \beta^2$ is left *unchanged*, i.e., it is *invariant* under $\mathbf{F}(v)$. As demonstrated in the next section, the non-Euclidean character of the internal 2-space also ensures that certain scalar-products of vectors such as $\mathbf{Q} \bullet \mathbf{Q} = \mathbf{Q}^2$ are, necessarily, C-reversing "charges" (not probability densities!).

Clearly, **g** is a C-reversing matrix since $(-\boldsymbol{\sigma}_x)\mathbf{g}(-\boldsymbol{\sigma}_x) = -\mathbf{g}$, which means that the components of **g** are also C-reversing "charges." What are these charges? Since $\mathbf{g} = \mathbf{F}_{\text{diag}}$, the components of **g** can be interpreted to be the scalar fermion-numbers, i.e., $g_{11} = f_m$, $g_{22} = f_a$, $g_{12} = g_{21} = 0$. Similar statements apply to the matrix $\boldsymbol{\sigma}_z$ [see Eq. (1.23)] since $\mathbf{g} = \mathbf{F}_{\text{diag}} = \boldsymbol{\sigma}_z$.

1.2. A NEW ORGANIZING PRINCIPLE

1.2.5 The Metric and Scalar-Products

Given the non-Euclidean metric in (1.26) and Eqs. (1.18) and (1.19), it follows that (nonzero) *scalar-products* of the eigenvectors of **F** are, necessarily, C-reversing charge-like quantities. In particular, the *squares* of **Q** and **Q**^{c} are C-reversing since

$$\mathbf{Q}^2 = q_1^2 - q_2^2 \tag{1.27}$$

and

$$(\mathbf{Q}^c)^2 = (-q_2)^2 - (-q_1)^2 \tag{1.28}$$

or

$$\mathbf{Q}^2 = -(\mathbf{Q}^c)^2. \tag{1.29}$$

In summary, *the scalar components of* **F**, **Q**, **Q**^{c} *and scalar-products such as* **Q**^{2} *and* (**Q**^{c})^2 *are all C-reversing dimensionless charge-like scalar quantities.* And, just as the scalar f is "carried" by an individual "isolated" fundamental-fermion or antifermion in a spacetime setting, I assume that all of these internal 2-scalars and their associated 2-space are, in some abstract sense, "carried" by individual "isolated" fundamental-fermions or antifermions in a spacetime setting (i.e., they are also Lorentz 4-scalars). Moreover, scalar-products and scalar-components associated with other vectors in the 2-space will be identified with additional *internal*, global charge-like quantum observables ("charges") carried by fundamental fermions. Certain mutually-commuting (compatible) sets of the various (conserved) scalar "charges" will serve as "good" quantum numbers that will be used to define simultaneous flavor-eigenstates (see Ref. 25, pp. 52 and 115).

1.2.6 Restatement of the New Organizing Principle

Given the 2-space metric in (1.26), and anticipating that the parameter v in (1.21) provides a natural way to incorporate the quark-lepton dichotomy, the new organizing principle for fundamental fermions will now be restated in its final form:

Individual fundamental-fermions ($q =$ *quarks,* $\ell =$ *leptons), and their associated flavor-doublets, can be described by "generalized fermion-numbers" consisting of* 2×2 *real matrices* **F** [**F** $=$ **F**(v) *for quarks and* **F** $=$ **F**(v') *for* *leptons*]. *These transformation matrices "act" on an internal, real two-dimensional non-Euclidean vector-space through an eigenvalue equation* $\mathbf{FQ} = f_m \mathbf{Q}$ *or* $\mathbf{FQ}^c = f_a \mathbf{Q}^c$, *where the scalar fermion-number* $f_m = +1$ *describes matter* ($\mathbf{Q} = \mathbf{Q}_q$ *or* $\mathbf{Q}_\ell$) *and* $f_a = -1$ *describes antimatter* ($\mathbf{Q}^c = \mathbf{Q}_q^c$ *or* $\mathbf{Q}_\ell^c$). *Specifically,* $\mathbf{Q}(\mathbf{Q}^c)$ *is a real, fixed 2-vector that represents both individual fundamental-fermions (antifermions) and their associated "up"-"down"-type flavor-doublets (antiflavor-doublets).*

Note that three quantum-dichotomies associated with fundamental fermions are essentially "built-in" to the new organizing principle. Consider, for example, the first-family of fundamental fermions. The dichotomies in question are the "up"-"down"-type flavor dichotomy associated with weak interactions and $SU(2)_L$ (e.g., $u - d$, $\nu_e - e^-$, etc.), the matter-antimatter dichotomy associated with quantum mechanics, special relativity and **PCT** (e.g., $u - \bar{u}$, $e^- - e^+$, etc.) and the quark-lepton "dichotomy" or, more accurately, the "strong-electroweak" dichotomy associated with strong interactions and $SU(3)_c$ (e.g., $u - e^-$, $d - \nu_e$, etc.). The latter dichotomy distinguishes electroweakly-interacting quarks, which are also *strongly-interacting* (quarks are strong-color triplets), from electroweakly-interacting leptons, which do not exhibit strong interactions[9] (leptons are strong-color singlets). For a conventional description of these dichotomies in the standard model or a five-color GUT such as $SU(5)$, see Refs. 4 and 5 and Appendix A.

While it is not immediately obvious, the new organizing principle not only incorporates three quantum-dichotomies, but it also provides the framework for an explanation of family *replication* and the *number* of families. The essence of the explanation, which will be "fleshed out" in more detail later, follows.

By the definition of a linear-vector 2-space, a 2-vector such as $\mathbf{Q}$ can always be resolved into a pair (no more, or less) of linearly-independent vectors $\mathbf{U}$ and $\mathbf{V}$ as $\mathbf{Q} = \mathbf{U} + \mathbf{V}$ (see Ref. 20, p. 31). And, according to the new organizing principle, since $\mathbf{Q}$ *represents* a flavor doublet, so should $\mathbf{U}$ and $\mathbf{V}$ *represent* this *same* flavor doublet. But, if this is so, *different vector-resolutions of* $\mathbf{Q}$ *should correspond to different flavor-doublets having the same* $\mathbf{Q}$. In other words, flavor doublets should be *replicated*.

Since $\mathbf{Q}$ can be resolved (mathematically) in an infinite number of ways, we might suppose that there are an *infinite* number of flavor doublets and hence families. But, because of various "quantum constraints," we will be able to show that $\mathbf{Q}$ *can be resolved in only three physically acceptable ways* *for **Q**-vectors associated with either quarks or leptons.* In other words, there can be only six quark flavors and six lepton flavors, which leads to the (*ex post facto*) "prediction" of three quark-lepton families.

Connections between the new organizing principle and the standard model or $SU(5)$ pictures, including connections to conventional internal color-charges, will be elucidated later in the book. Meanwhile, the physical meaning of certain geometric objects appearing in (or implied by) the new organizing principle will be explored further in the next section.

1.3 Preliminary Physical Interpretation

The internal properties of elementary particles are traditionally described by non-abelian (non-commutative) Lie algebras $SU(N)$. These algebras are associated with Euclidean vector-spaces having $N^2 - 1$ dimensions and an equal number of generators, $N-1$ of which constitute a Cartan subalgebra of mutually-commuting *diagonal* generators.[26] For example, the simplest possible non-abelian Lie algebra is $SU(2)$, which is associated with a Euclidean 3-space and has just one *diagonal* generator λ_3.

The proposed 2-space cannot be associated directly with a non-abelian Lie algebra since all of these algebras have associated vector-spaces of three, or more, dimensions. Hence, it is possible, as the previous discussions have suggested, that the proposed 2-space is associated with a new type of internal abelian (commutative) algebraic structure superficially resembling a Cartan subalgebra, and involving a number of different conserved mutually-commuting $U(1)$-type charges [e.g., baryon number B is a $U(1)$-type charge that generates a phase transformation $\psi \to e^{-i\alpha B}\psi$].

The explanation for these $U(1)$-type charges may lie at the Planck level where all flavor degrees of freedom presumably originate. But, no detailed attempt will be made in this book to explain the new organizing principle and its associated charges in terms of deeper principles or physics at the Planck level. Instead, this book will be devoted almost exclusively to working out the algebraic, geometric and physical consequences of the new organizing principle as if it were known to be true.

1.3.1 Mutually-Commuting Global Charges and the 2-Space

As previously emphasized, the eigenvalue equations (1.12) and (1.13) involving $\mathbf{F}$ and Eqs. (1.1) and (1.2), respectively, involving $\mathbf{F}(\text{op})$, have a similar appearance but represent entirely different physical situations—even though they both describe the matter-antimatter dichotomy. $\mathbf{F}$ is *not* a quantum operator and the eigenvectors $\mathbf{Q}$ and $\mathbf{Q}^c$ are *not* single-particle quantum state-vectors in Hilbert space. Instead, the vectors $\mathbf{Q}$ or $\mathbf{Q}^c$ are taken to reside in the new *internal* 2-space, where they represent some *intrinsic* "physical property" of *individual* fundamental-fermions or antifermions (and their associated flavor-doublets), respectively. In this very general sense, the vectors $\mathbf{Q}$ and $\mathbf{Q}^c$ are analogous to the total-isospin vector $\mathbf{T}$, or the total angular-momentum vector $\mathbf{J}$, of $SU(2)$. But, there are several very important ways in which these 2-vectors differ from a vector like $\mathbf{T}$ or $\mathbf{J}$.

First, the space of $\mathbf{T}$ is a three-dimensional Euclidean space instead of a two-dimensional non-Euclidean space. Second, the total-isospin square, namely, $\mathbf{T}^2$ (like $\mathbf{J}^2$) is a C-invariant *mass-like* scalar, whereas $\mathbf{Q}^2$ is a C-reversing *charge* [see Eq. (1.29)]. Third, just as for the total angular-momentum vector $\mathbf{J}$ in ordinary 3-space, in the 3-space associated with $\mathbf{T}$ the uncertainty principle, together with the noncommutativity of rotations in any Euclidean 3-space, *precludes* simultaneous knowledge of the three vector-components T_x, T_y and T_z (see Ref. 25, p. 144). In particular, because the quantum *operators* that must be used to represent these components do *not* commute (i.e., they are *incompatible*) a measurement of one component of $\mathbf{T}$ always interferes with the measurement of its other components at a later time.

This kind of quantum interference must not occur in the 2-space. That is, if $\mathbf{F}$ is to be a generally non-Hermitian real matrix that possesses real eigenvalues and real eigenvectors (a well-defined eigenvalue equation), and if all of these things are to be intrinsic, physically-observable properties of individual "isolated" fundamental-fermions, then: *every component of $\mathbf{F}$ and every component of its associated eigenvectors (and its eigenvalues) must be known, in principle, precisely and simultaneously.* Otherwise, the proposed generalization ($f \Rightarrow \mathbf{F}$) and Eqs. (1.3) through (1.9), which embody it, would be both mathematically inappropriate and physically meaningless. Hence, we must assume that all of the observable physical quantities (i.e., "charges") represented by these 2-space components are consistent with the requirement that they are compatible and *mutually commute*. Under these conditions, a measurement of any one of these 2-space components will *not* interfere with

1.3. PRELIMINARY PHYSICAL INTERPRETATION

the measurement of any other component at a later time.

Finally, because **Q** represents *both* individual *flavors* and *flavor-doublets* the charge **Q**^{2}—like any C-reversing charge—is additive for a "composite" of two different fundamental-fermions each described by **Q**. But, their corresponding **Q**-vectors do *not* "interfere," i.e., they cannot be superimposed. If they could, a self-consistent physical picture would require a "nonsensical" charge, namely, $(\mathbf{Q}+\mathbf{Q})^2 = 4\mathbf{Q}^2$ to be one of the "charges" carried by the "composite" in addition to the "sensible" charge $\mathbf{Q}^2 + \mathbf{Q}^2 = 2\mathbf{Q}^2$, as is conventionally required for additive charges. Therefore, the respective vectors **Q**, **U** or **V** for two *different* fundamental-fermions cannot be superimposed. But, their respective charges (e.g., the components of **Q**, **U** or **V** and/or the scalars $\mathbf{Q}^2$, $\mathbf{U}^2$, $\mathbf{U} \bullet \mathbf{V}$ and $\mathbf{V}^2$) can be superimposed (combined).

In a loose (algebraic) sense this means that *the 2-spaces associated with (carried by) two different fundamental-fermions are effectively "isolated" from one another.* This situation is roughly the converse of the situation for a vector such as **T**. The C-invariant scalar $\mathbf{T}^2$, for each of two different particles carrying **T**-vectors, is not additive for a "composite", but the corresponding **T**-vectors are, i.e., they interfere or interact with each other in isospin space.

The foregoing conserved mutually-commuting (compatible) 2-space "charges" are reminiscent of a Cartan subalgebra[26], and they may have a similar utility in defining flavor eigenstates. Recall, for example, that $SU(3)$-flavor has two mutually-commuting diagonal generators (λ_3 and λ_8), which leads to the well-known Gell-Mann–Nishijima formula for the quark electric-charges, namely,[26,27]

$$Q = T_3 + Y/2. \tag{1.30}$$

In this formula, T_3 is the third-component of (global) isospin,

$$Y = B + S \tag{1.31}$$

is the (global) hypercharge, B is the baryon number and S is the strangeness. And, because the conserved "charges" T_3 and Y are represented by operators that are proportional to the mutually-commuting diagonal generators λ_3 and λ_8, respectively, they are "good" $U(1)$-type quantum numbers (in strong and electromagnetic interactions) that can be used to define *simultaneous flavor-eigenstates* of the form (see Ref. 25, pp. 52 and 115 and Ref. 26)

$$|T_3, Y\rangle.$$

These eigenstates can be used to represent the u, d and s quark flavors. In summary, the mutually-commuting 2-space charges dealt with in this book are imagined to be very similar to the foregoing "diagonal" charges of $SU(3)$-flavor, i.e., they are assumed to be "good" (conserved) quantum numbers that can be used to express the quark and lepton electric-charges and simultaneous flavor-eigenstates.

1.4 Representing Flavor Eigenstates and Flavor Doublets

Given that the scalar components of vectors such as **Q** and **Q**^{c}, and the scalar components of matrices such as **F**, all mutually-commute and transform like C-reversing *charges*—we should be able to represent simultaneous flavor-eigenstates by assembling an appropriate collection of these conserved flavor-defining "charges" or "good quantum numbers." To proceed with this program we must find all such charges and then find a way to *quantize* them. First, let us find these charges.

Because the space is two-dimensional, the fixed (constant) eigenvectors of $\mathbf{F}(v)$ and $\mathbf{F}(v')$ that describe quark ($\mathbf{Q}_q$) and lepton ($\mathbf{Q}_\ell$) flavors and flavor-doublets, respectively, can be resolved into appropriate pairs of linearly-independent "basis" vectors **U** and **V** (see Ref. 20, p. 31). And, because the scalar components of the vectors **Q** are charge-like, the scalar components of every vector associated with a vector resolution of **Q**, such as

$$\mathbf{Q} = \mathbf{U} + \mathbf{V}, \qquad (1.32)$$

should be charge-like. Finally, every non-Euclidean scalar-product of vectors appearing in an equation, such as the "square" of (1.32), namely,

$$\mathbf{Q}^2 = \mathbf{U}^2 + 2\mathbf{U} \bullet \mathbf{V} + \mathbf{V}^2, \qquad (1.33)$$

must be charge-like [see Eq. (1.29)]. I now assume that *the flavor-defining "charges," or charge-like quantum numbers, defined by Eqs. (1.32) and (1.33) exhausts the possibilities and that these charges suffice to define the fundamental-fermion simultaneous flavor-eigenstates.*[28] Once again, we should emphasize that, except for the electric charge, which is both a locally- and a globally-conserved charge, these charges, like f, are assumed to be strictly global. Moreover, like f, these "charges" are taken to be 2-scalars in the *internal* 2-space and conserved 4-scalars in *spacetime*.

1.4. FLAVOR EIGENSTATES AND FLAVOR DOUBLETS

1.4.1 Flavor Doublets

Since only *one* fixed (constant) charge-vector $\mathbf{Q}_q(\mathbf{Q}_\ell)$ is available to describe a multiplicity of otherwise very *different* quark (lepton) flavor-doublets (see Sec. 1.2.3), I now assume:

1. The *two* charge-like components of $\mathbf{Q}_q(\mathbf{Q}_\ell)$ are quark (lepton) *electric* charges. By definition, these and other charge-like vector components are "measured" along one of four *different*, positive or negative, 2-space coordinate-axis directions [see charge-label subscripts 1, 2 in (1.18)].
2. There is a one-to-one correspondence between charge-vector "triads" $(\mathbf{Q}_q, \mathbf{U}, \mathbf{V})$ or $(\mathbf{Q}_\ell, \mathbf{U}, \mathbf{V})$ and individual flavors or flavor doublets, where $\mathbf{Q}_q = \mathbf{U} + \mathbf{V}$ and $\mathbf{Q}_\ell = \mathbf{U} + \mathbf{V}$, and $\mathbf{U}$ and $\mathbf{V}$ are different for different flavors or flavor doublets (see Sec. 1.2.6).

The general quark "basis" vectors, in column-vector form, are

$$\mathbf{U} = \{u_1, u_2\}$$

and

$$\mathbf{V} = \{v_1, v_2\},$$

where u_1, u_2, v_1 and v_2 are a "minimal" set of *independent* charge-like quantum numbers. Similarly, the charge-like quantum numbers u'_1, u'_2, v'_1, v'_2 are used to define **U**- and **V**-vectors for the leptons.[29] Finally, the column vectors

$$\mathbf{Q}_q = \{q_1, q_2\} \quad (1.34)$$

and

$$\mathbf{Q}_\ell = \{q'_1, q'_2\}, \quad (1.35)$$

which are resolved by **U** and **V**, describe quarks and leptons, respectively.

Using (1.34) and **U**- and **V**-vectors appropriate to quarks, the two quark flavors associated with the respective "up"-"down"-type electric charges q_1 or q_2 will be *represented* by simultaneous flavor-eigenstates of the general form (other forms such as $|u_1, v_1\rangle$ and $|u_2, v_2\rangle$ will also be used):

$$|q_1, u_1, v_1, \mathbf{Q}_q^2, \mathbf{U}^2, 2\mathbf{U} \bullet \mathbf{V}, \mathbf{V}^2\rangle$$
$$|q_2, u_2, v_2, \mathbf{Q}_q^2, \mathbf{U}^2, 2\mathbf{U} \bullet \mathbf{V}, \mathbf{V}^2\rangle,$$

where the various global charges are related as follows:

$$q_1 = u_1 + v_1$$
$$q_2 = u_2 + v_2$$
$$\mathbf{Q}_q^2 = q_1^2 - q_2^2$$
$$\mathbf{U}^2 = u_1^2 - u_2^2$$
$$2\mathbf{U} \bullet \mathbf{V} = 2(u_1 v_1 - u_2 v_2)$$
$$\mathbf{V}^2 = v_1^2 - v_2^2.$$

Algebraically speaking, at least, the *dependent* variables $\mathbf{Q}_q^2, \mathbf{U}^2, 2\mathbf{U} \bullet \mathbf{V}$ and $\mathbf{V}^2$ are "less" fundamental than the *independent* variables u_1, u_2, v_1, v_2.

The **C**-reversed states, and the lepton flavor-states, can be treated in exactly the same way, provided appropriate **U**- and **V**-vectors are used. Thus, it takes *four* scalar charge-labels (i.e., the charge-like quantum numbers u_1, u_2, v_1, and v_2) to specify each of two flavors in any given quark flavor-doublet. For example, it will be shown for the u, d and s quarks that $\mathbf{Q}_q^2 = B$ is the baryon number, $\mathbf{S} = -\mathbf{U}^2$ is the strangeness, $Y = 2\mathbf{U} \bullet \mathbf{V}$ is the (global) hypercharge, and the components of **U** (for the u and d quarks) are the third-components[29] of (global) isospin, namely, T_3. Similar statements apply to lepton flavor-doublets. Hence, without additional constraints, it takes a "minimal" set of eight independent quantum-numbers (i.e., $u_1, u_2, v_1, v_2, u_1', u_2', v_1'$ and v_2') to specify the four flavors in any quark-lepton family.[28] Clearly, because **U** and **V** may be chosen in *different* ways, there will be more than one family.

1.4.2 Flavor Conservation

Because we are assuming that all of the flavor-defining 2-scalar charge labels identified in Sec. 1.4.1 are "good" quantum numbers, they must be constants of the motion (conserved) for "*isolated*" fermion or antifermion flavors. But, even in the presence of interactions, we expect these scalars to exhibit a strong tendency to be separately conserved. Accordingly, I assume that these quantum numbers are *exactly* conserved in all *strong, electromagnetic* and *weak neutral-current interactions* and, therefore, that flavors are conserved by these interactions. In weak charge-changing interactions some of these flavor-defining charges will cease to be "good" quantum numbers. And, except for electric charge, *all* of these global charges will be destroyed in gravitational collapse.

1.4. FLAVOR EIGENSTATES AND FLAVOR DOUBLETS

1.4.3 Effective "Quantization"

The foregoing proposal is only a framework for a description of fundamental fermions. There are only hints in what has been presented thus far to indicate how the components of vectors like **Q** (i.e., electric charges) are to be *quantized*.

Quantization would be achieved if some way could be found to limit the number of **U**- and **V**-vectors or, equivalently, the number of **Q**- and **U**- or **Q**- and **V**-vectors. Then, the number of flavors, flavor-doublets and families would be limited because there would be a limited number of "vector-triads" (**Q**, **U**, **V**) available to represent individual flavors and flavor-doublets (see Sec. 1.4.1).

Let us consider how to find all possible **Q**- and **U**-vectors. Assumption (1) in Sec. 1.4.1 means that the components of vectors like **Q**, being the electric charges of the fundamental fermions, already obey well-known laws of quantization. These laws depend, in turn, on internal color-charges [e.g., $SU(3)_c$ and/or $SU(5)$ color-charges].

Now, since **Q** is also an eigenvector of $\mathbf{F}(v)$, the 2-vector **Q** and its components (the electric charges carried by fundamental fermions) are all functions of the dimensionless-parameter v (see Sec. 1.2.3). Similarly, the baryon number $\mathbf{Q}_q^2 = B$ is a function of v. In this way, a connection between electric charge, internal color-charges, and the parameter v can be made. In particular, we will be able to show that v obeys the "quantum condition":

$$v = \ln M_c, \qquad (1.36)$$

where M_c counts both the number of fundamental fermions in a strongly-bound composite fermion, and the strong-color multiplicity. That is, we will show that M_c in (1.36) is $M_c = 3$ for quarks (strong-color triplets) and $M_c = 1$ for leptons (strong-color singlets). Finally, without elaboration here, the possible **U**-vectors will be found by requiring their components (u_1 and u_2) to be **C**-reversing "charges" analogous to the third-component of global isospin T_3, which serve to distinguish "up"- and "down"-type flavors.

Remarkably, only six **U**-vectors exist that provide such distinctions (three for quarks and three for leptons). In this way, we are able to show ("predict" *ex post facto*) that only six types of flavor doublets and three types of quark-lepton families can be given a self-consistent description in the present scheme.

In a sense, if you have an elementary particle, and you describe how it behaves under various symmetry transformations, including translations, rotations, gauge transformations, then you've said everything there is to say about the particle.
—S. Weinberg

2. The 2-Space Description of First-Family Fermions

The framework for describing quarks and leptons proposed in Chapter 1 will now be applied to the first family. By expressing known first-family particle "charges" (e.g., electric charge Q, baryon number B and lepton number L) as scalars in the new 2-space, we will learn how to identify and interpret these and other more general *flavor-defining* charge labels. Eventually, this process will lead to the incorporation of flavors and flavor-doublets beyond those in the first family.

In carrying out this program, I will take as given both the experimentally determined first-family particle properties (e.g., Q, B, L, etc.) and their accepted theoretical descriptions within the standard model[4] $SU(3)_c \times SU(2)_L \times U(1)_Y$ and/or GUTs such as $SU(5)$.[4-9] In particular, I will assume that the proposed 2-space description of flavors, flavor-doublets and families is, in some approximate sense, complementary to, and compatible with, these conventional theoretical descriptions.

2.1 Family Definition and the Family Hierarchy

Quark and lepton flavor, or mass, eigenstates are known to fall in a hierarchy of repeated family-associations exhibiting increasing average mass. The first such family of fundamental fermions consists of the following combinations of left-handed (L) flavor-doublet and right-handed (R) flavor-singlet particles, namely, Family I (least average mass):

$$\begin{pmatrix} \nu_e \\ e^- \end{pmatrix}_L \begin{pmatrix} u \\ d \end{pmatrix}_L e^-_R u_R d_R.$$

The remaining two families, which we will later *infer* by a process of generalization from the first family are, in order of increasing average mass,

Family II (intermediate average mass):

$$\begin{pmatrix} \nu_\mu \\ \mu^- \end{pmatrix}_L \begin{pmatrix} c \\ s \end{pmatrix}_L \mu_R^- c_R s_R,$$

and Family III (greatest average mass):

$$\begin{pmatrix} \nu_\tau \\ \tau^- \end{pmatrix}_L \begin{pmatrix} t \\ b \end{pmatrix}_L \tau_R^- t_R b_R.$$

For the purposes of this book we will focus attention on left-handed fermions and right-handed antifermions only, since only these particles are nonsinglets under weak interactions.[4] Hence, the first "family" will be defined hereinafter to consist of a set of four left-handed fermion flavor, or mass, eigenstates (u, d, e^-, ν_e) made up of an ("up," "down")-type left-handed quark flavor-doublet (u, d), and a ("down," "up")-type left-handed lepton flavor-doublet (e^-, ν_e). And, even though the subscript L does not appear in these (in-text) definitions it is to be understood.

Because each quark flavor is a strong-color triplet, and each lepton flavor is a strong-color singlet, there are a total of eight left-handed states in the first family. Similarly, the other two families, namely, the second family (c, s, μ^-, ν_μ) and the third family (t, b, τ^-, ν_τ) consist of eight left-handed states.

2.2 Labeling and Subscript Conventions

In Chapter 3, we will further elucidate the 2-space geometry and topology by representing the non-Euclidean 2-space on a Euclidean plane ($E2$). To facilitate this presentation it is necessary to define certain coordinate axes and certain coordinate-axis directions in the 2-space.

Because the electric charges of fundamental fermions are associated with the 2-vectors **Q** (see Assumption 1 in Sec. 1.4.1) they are, necessarily, "measured" along one of *two* possible coordinate-axes in the 2-space. Accordingly, the electric charges (and other charges) for material particles in the first family will be labeled with "up"-"down"-type subscripts (i.e., 1 or 2) that refer to certain orthogonal coordinate-axes in the 2-space. These will be called the X_1 or X_2 axes, respectively. Thus, *the four electric-charges associated with the four first-family flavors will be used to define the internal 2-space.*

By definition, the X_1 and X_2 coordinate-axes correspond, respectively, to the two one-dimensional *invariant vector spaces*[30] associated with the matrices $+\mathbf{F}_{\text{diag}}$ and $-\mathbf{F}_{\text{diag}}$, where $\mathbf{F}_{\text{diag}} = \mathbf{g} = \sigma_z$ [see Eqs. (1.5), (1.23) and (1.26)]. That is, these two subspaces are defined by an infinite number of eigenvectors associated with $\pm \mathbf{F}_{\text{diag}}$, each having the eigenvalue $+1$, namely, the vectors $\{k_1, 0\}$ along the X_1 axis and the vectors $\{0, k_2\}$ along the X_2 axis, where k_1 and k_2 are arbitrary real numbers.

Using these axes as a reference, the $u(d)$ quark carrying baryon number $B = +1/3$ is *assigned* the electric charge designated $q_1 = +2/3$ ($q_2 = -1/3$), while the e^- (ν_e) lepton carrying lepton number $L = +1$ is *assigned* the electric charge designated $q'_1 = -1$ ($q'_2 = 0$). That is, the *four* electric-charges known to be associated with matter, namely, $q_1 = +2/3$, $q_2 = -1/3$, $q'_1 = -1$ and $q'_2 = 0$, which include charges of zero magnitude, *are taken to be* "measured" along the four different 2-space coordinate-axis directions $+X_1$, $-X_2$, $-X_1$, and $+X_2$, respectively.

Similarly, the **C**-reversed first-family counterparts, namely, the $\bar{u}$ ($\bar{d}$) anti-quark carrying baryon number $B = -1/3$ is *assigned* the **C**-reversed electric-charge [see Eq. (1.19)] designated $-q_1 = -2/3$ ($-q_2 = +1/3$), while the e^+($\bar{\nu}_e$) antilepton carrying lepton number $L = -1$ is *assigned* the **C**-reversed electric-charge designated $-q'_1 = +1$ ($-q'_2 = 0$).[31] Owing to the fact that both the order, and the signs, of 2-vector components are reversed by $-\sigma_x$, the *four* electric-charges associated with antimatter, namely, $-q_1 = -2/3$, $-q_2 = +1/3$, $-q'_1 = +1$ and $-q'_2 = 0$ *are, necessarily, "measured" along the four coordinate-axis directions,* $-X_2$, $+X_1$, $+X_2$ and $-X_1$, respectively.

In Table I the foregoing assignments and associations are tabulated for future reference. It should be noted that B and L in Table I are expressed as scalar-products involving 2-vectors of the type described by Eqs. (1.18) and (1.19). This step will be justified in the next section.

2.3 B and L as Functions of Electric Charge Q

Instead of starting with the "generalized fermion-number" $\mathbf{F}$, and then attempting to work out such things as B and L, I will begin by expressing B and L directly in terms of 2-space electric-charge coordinates. This step will be justified both by the 2-space algebra and geometry and by a GUT such as $SU(5)$.

In "diagonal" weak-transitions of quarks or leptons, initial-state "up" ("down")-type particles are transformed into final-state "down" ("up")-type

2.3. B AND L AS FUNCTIONS OF ELECTRIC CHARGE Q

Table I. Labeling and Subscript Conventions for First-Family Fermions and Antifermions. The electric charges Q of these particles are "measured" along one of two orthogonal coordinate-axes, namely, the X_1 or X_2 axes. By definition, the components a and b of *any* 2-vector describing matter (e.g., $\mathbf{Q}$, $\mathbf{U}$ or $\mathbf{V}$), either in column-form $\{a, b\}$ or in row-form (a, b), are "measured" along the X_1 and X_2 coordinate axes, respectively. Baryon number B, lepton number L, and the specific "measurement direction" (i.e., $\pm X_1$ or $\pm X_2$) for the electric charges Q, is indicated. The one-to-one correspondence between flavor doublets and certain 2-vectors such as $\mathbf{Q}_q = \{q_1, q_2\} = \{2/3, -1/3\}$ or $\mathbf{Q}_\ell = \{q'_1, q'_2\} = \{-1, 0\}$, is indicated at the bottom of the table. The electric charge Q of a particular quark or lepton flavor is the Euclidean-projection of the 2-vector $\mathbf{Q}_q = \{q_1, q_2\}$ or $\mathbf{Q}_\ell = \{q'_1, q'_2\}$, respectively, on one of the two coordinate-axes X_1 or X_2. We may also think of the charge Q as being a non-Euclidean "projection" defined by the following non-Euclidean scalar-products for quarks: $q_1 = (1,0)\{\frac{2}{3}, -\frac{1}{3}\} = \frac{2}{3}$, $q_2 = (0,-1)\{\frac{2}{3}, -\frac{1}{3}\} = -\frac{1}{3}$, $-q_1 = (0,-1)\{\frac{1}{3}, -\frac{2}{3}\} = -\frac{2}{3}$, and $-q_2 = (1,0)\{\frac{1}{3}, -\frac{2}{3}\} = \frac{1}{3}$, where the vectors $(0,-1)$ and $(1,0)$ are directed from the origin along the X_2 or X_1 coordinate axes, respectively. Similar statements apply to the leptons and to the vectors $\mathbf{U}$ and $\mathbf{V}$.

Flavor	$B = \mathbf{Q}_q^2, (\mathbf{Q}_q^c)^2$	$L = \mathbf{Q}_\ell^2, (\mathbf{Q}_\ell^c)^2$	Q	Measurement direction	Quadrant
u	$+1/3$	0	$q_1 = +2/3$	$+X_1$	
d	$+1/3$	0	$q_2 = -1/3$	$-X_2$	IV
$\bar{d}$	$-1/3$	0	$-q_2 = +1/3$	$+X_1$	
$\bar{u}$	$-1/3$	0	$-q_1 = -2/3$	$-X_2$	
ν_e	0	$+1$	$q'_2 = 0$	$+X_2$	
e^-	0	$+1$	$q'_1 = -1$	$-X_1$	II
e^+	0	-1	$-q'_1 = +1$	$+X_2$	
$\bar{\nu}_e$	0	-1	$-q'_2 = 0$	$-X_1$	

$$(u, d) \Leftrightarrow \mathbf{Q}_q = \{q_1, q_2\}; \quad (\bar{d}, \bar{u}) \Leftrightarrow \mathbf{Q}_q^c = \{-q_2, -q_1\}$$
$$(e^-, \nu_e) \Leftrightarrow \mathbf{Q}_\ell = \{q'_1, q'_2\}; \quad (\bar{\nu}_e, e^+) \Leftrightarrow \mathbf{Q}_\ell^c = \{-q'_2, -q'_1\}$$

particles within the same flavor-doublet. And, while the quark or lepton electric-charge always *changes* by unity in such a process (e.g., $u \to d + W^+$), the baryon number B or lepton number $B' = L$ is always *unchanged*. This suggests that B and L can be represented as some invariant function of "up"- and "down"-type electric charges.

2.3.1 B and L as Linear Functions of Electric Charge Q

In $SU(5)$ the internal color-charge content of the eight left-handed first-family *fermions* determines their "up"-"down"-type electric-charges Q through the relation (see Footnote 4, Ref. 5 and Appendix A)

$$Q = G/2 + Y(\text{weak})/2. \tag{2.1}$$

Here, G stands for "green", a weak-color charge [$G/2$ is the third-component of *weak-isospin* T_3(weak) for left-handed fermions or right-handed antifermions only] and Y(weak) is the *weak-hypercharge*, which depends on the three strong-color charges.

Now, Y(weak) also equals twice the arithmetic average of the "up" and "down" electric-charges in a flavor doublet, namely, the *sum* of the "up" and "down" electric charges. Finally, B or L for left-handed fermions or right-handed antifermions can be written simply as $B = Y$(weak) or $L = -Y$(weak). Therefore, in $SU(5)$ or *any* other description that correctly yields the experimentally observed electric charges of fundamental fermions, B and L may always be expressed as being proportional to the sum of the "up"- and "down" type electric charge labels in a flavor doublet. But, it will also be shown below that it is equally natural (in the 2-space) to represent B and L by *quadratic* functions of these same "up"- and "down"-type electric charges.

2.3.2   General B and L Values as Quadratic Functions of Q

Using the labeling conventions of Table I, and noting that "up"- and "down"-type C-reversing electric charges differ by the fundamental unit of electric charge (i.e., $G = \pm 1$ or $q_1 - q_2 = +1$ and $q'_1 - q'_2 = -1$), it follows that B and L may always be expressed, either as a linear combination of "up"- and "down"-type electric-charges (see Sec. 2.3.1), or as a *quadratic* in these same charges. For the first family of left-handed fundamental fermions (u, d, e^-, ν_e)

$$B = q_1 + q_2 = q_1^2 - q_2^2 = +Y(\text{weak}) = +1/3 \tag{2.2}$$

2.4. SCALAR-PRODUCTS OF 2-VECTORS

and

$$L = -q_1' - q_2' = (q_1')^2 - (q_2')^2 = -Y(\text{weak}) = +1. \tag{2.3}$$

Similarly, for the corresponding C-reversed right-handed antifermion family $(\bar{d}, \bar{u}, \bar{\nu}_e, e^+)$ we have

$$B = -q_2 - q_1 = (-q_2)^2 - (-q_1)^2 = +Y(\text{weak}) = -1/3 \tag{2.4}$$

and

$$L = q_2' + q_1' = (-q_2')^2 - (-q_1')^2 = -Y(\text{weak}) = -1. \tag{2.5}$$

I now assume that quark (lepton) flavor-doublets carry these values of B (L) no matter in which family they reside. Evidently, this assumption is equivalent to the assumption that all flavor doublets are associated with the *same* electric charges.

2.4 Scalar-Products of 2-Vectors

The labeling conventions of Table I permit each of the quadratic expressions in Eqs. (2.2) through (2.5) to be represented as a *scalar-product* of 2-vectors. Using the non-Euclidean metric given by (1.26), the baryon number B for both members of any given quark flavor-doublet can be written as

$$B = \Sigma g_{ij} q_i q_j, \tag{2.6}$$

where g_{ij} are the components of the metric, and the summation indicated is over repeated indices or subscripts, where $i, j = 1$ to 2. Expanding Eqs. (2.6) using (1.26), the result is

$$B = g_{11} q_1 q_1 + g_{12} q_1 q_2 + g_{21} q_2 q_1 + g_{22} q_2 q_2 = q_1^2 - q_2^2, \tag{2.7}$$

which is the same as the quadratic in (2.2), as required.

The baryon number B can also be written in symbolic form as the non-Euclidean scalar-product of a *row* vector (q_1, q_2) and a conformable *column* vector $\{q_1, q_2\}$, namely,

$$B = (q_1, q_2)\{q_1, q_2\} = \{q_1, q_2\}^2,$$

where $\{q_1, q_2\}^2$ is "shorthand" notation for the "square" of the vector. Sometimes non-Euclidean scalar-products will be written using a "dot" interposed between vectors (represented by bold capital letters), namely, $\mathbf{Q} \bullet \mathbf{Q} = \mathbf{Q}^2$ and/or

$$\mathbf{Q} \bullet \mathbf{Q} = (q_1, q_2)\{q_1, q_2\}.$$

Similar statements apply to the vectors $\mathbf{U}$ and $\mathbf{V}$.

In general, the non-Euclidean scalar-product of any 2-vector (x_1, y_1) with a conformable 2-vector $\{x_2, y_2\}$, yields an expression of the form (see Appendix B)

$$(x_1, y_1)\{x_2, y_2\} = x_1 x_2 - y_1 y_2,$$

which is *always* a charge-like scalar since it changes signs under the matter-antimatter transformation [e.g., see Eqs. (1.17) through (1.19)], namely,

$$(-y_1, -x_1)\{-y_2, -x_2\} = y_1 y_2 - x_1 x_2.$$

2.4.1 General B and L Values Expressed as Scalar-Products

If leptons and quarks are to be treated on an equal basis, both lepton-number L and baryon-number B should be expressed as scalar-products. In Table I a quark electric-charge vector written as

$$\mathbf{Q}_q = \{q_1, q_2\} = \{2/3, -1/3\}, \tag{2.8}$$

and its antiparticle counterpart written as

$$\mathbf{Q}_q^c = \{-q_2, -q_1\} = \{1/3, -2/3\}, \tag{2.9}$$

represent the first-family quark flavor-doublet (u, d) and the corresponding antiquark flavor-doublet $(\bar{d}, \bar{u})$, respectively. Similarly, the lepton electric-charge vector

$$\mathbf{Q}_\ell = \{q_1', q_2'\} = \{-1, 0\}, \tag{2.10}$$

and its antiparticle counterpart

$$\mathbf{Q}_\ell^c = \{-q_2', -q_1'\} = \{0, 1\}, \tag{2.11}$$

represent the first-family lepton flavor-doublet (e^-, ν_e) and the corresponding antilepton flavor-doublet $(\bar{\nu}_e, e^+)$, respectively.

2.4. SCALAR-PRODUCTS OF 2-VECTORS

Using the electric-charge vectors defined in Eqs. (2.8) through (2.11), and the metric **g** defined in (1.26), Eqs. (2.2) through (2.5) can be expressed as non-Euclidean scalar-products (i.e., as 2-scalars)

$$B = \mathbf{Q}_q^2 = (q_1, q_2)\{q_1, q_2\}$$

or

$$B = (\mathbf{Q}_q^c)^2 = (-q_2, -q_1)\{-q_2, -q_1\},$$

and

$$B' = L = \mathbf{Q}_\ell^2 = (q_1', q_2')\{q_1', q_2'\}$$

or

$$B' = L = (\mathbf{Q}_\ell^c)^2 = (-q_2', -q_1')\{-q_2', -q_1'\}.$$

(2.12)

2.4.2 Scalar Fermion-Numbers f and the Matrix F

The fact that B and L can be represented as 2-scalars in a two-dimensional non-Euclidean space (they are also taken to be Lorentz 4-scalars in space-time) means that the associated fermion-numbers f can similarly be represented as 2-scalars. In particular, since f can be thought of as a "normalized" baryon- or lepton-number, namely, $f = B/|B|$ or $f = L/|L|$, respectively, the scalar fermion-numbers for *quarks* and *antiquarks* can be written as (see p. 25)

$$f_m = \{q_1, q_2\}^2 / |\{q_1, q_2\}^2| = +1 \tag{2.13}$$

and

$$f_a = \{-q_2, -q_1\}^2 / |\{-q_2, -q_1\}^2| = -1, \tag{2.14}$$

respectively. Similar expressions apply to *leptons* and *antileptons*.

Evidently, the 2-scalars f_m and f_a, together with their respective 2-vectors $\{q_1, q_2\}$ and $\{-q_2, -q_1\}$, serve to *define* the 2×2 transformation matrix[21] **F** for quarks via the eigenvalue equations or transformation "laws"

$$\mathbf{F}\{q_1, q_2\} = f_m\{q_1, q_2\} \tag{2.15}$$

and
$$\mathbf{F}\{-q_2, -q_1\} = f_a\{-q_2, -q_1\}. \tag{2.16}$$

Similar statements apply to the leptons. Therefore, the previous description of B and L is consistent with the properties of the "generalized fermion-number" $\mathbf{F}$ (worked out in Chapter 1) and the new organizing principle.

2.4.3 Orthogonality of Matter and Antimatter

Matter and antimatter flavor-doublets, as represented by their respective electric-charge vectors, are "orthogonal" since [use Eqs. (2.8) through (2.11) and the metric $\mathbf{g}$ given in (1.26)]

$$\mathbf{Q}_q \bullet \mathbf{Q}_q^c = 0 \text{ or } \mathbf{Q}_\ell \bullet \mathbf{Q}_\ell^c = 0. \tag{2.17}$$

Similarly, $\mathbf{U} \bullet \mathbf{U}^c = \mathbf{V} \bullet \mathbf{V}^c = 0$ for *all* $\mathbf{U}$ and $\mathbf{V}$. The mathematical meaning of (2.17) is clear, but what is its physical meaning?

Temporarily dropping the subscripts q and ℓ on the electric-charge vectors, and using Eqs. (1.18), (1.19) and (1.26), it is easily demonstrated that

$$(\mathbf{Q} + \mathbf{Q}^c)^2 = 0,$$

where the superposition of $\mathbf{Q}$ and $\mathbf{Q}^c$ is only a *mathematical* one and does *not* mean that a fermion and an antifermion, or their respective flavor-doublets, are being superimposed in a physical sense (see Sec. 1.3.1). Then

$$\mathbf{Q}^2 + 2\mathbf{Q} \bullet \mathbf{Q}^c + (\mathbf{Q}^c)^2 = 0$$

and

$$\mathbf{Q} \bullet \mathbf{Q}^c = -1/2[\mathbf{Q}^2 + (\mathbf{Q}^c)^2].$$

But, from Sec. 2.4.2,

$$\mathbf{Q}^2 = |\mathbf{Q}^2| f_m,$$

and

$$(\mathbf{Q}^c)^2 = |(\mathbf{Q}^c)^2| f_a,$$

where

$$|\mathbf{Q}^2| = |(\mathbf{Q}^c)^2|.$$

Therefore,

$$\mathbf{Q} \bullet \mathbf{Q}^c = -1/2|\mathbf{Q}^2| (f_m + f_a),$$

or
$$\mathbf{Q} \bullet \mathbf{Q}^c = -1/2|\mathbf{Q}^2|\ tr\ \mathbf{F}.$$
Since $|\mathbf{Q}^2| \neq 0$ in general, if the fermion number f for *individual* fundamental-fermions is to be a C-reversing "charge" (i.e., $tr\ \mathbf{F} = 0$), the scalar-product $\mathbf{Q} \bullet \mathbf{Q}^c$ must vanish.

2.5 A Preliminary Geometric Interpretation of Three Quantum Dichotomies

As discussed in Chapter 1, the generalization of f to a *matrix* $\mathbf{F}$ automatically introduces new "degrees of freedom." For example, while 2-scalars f (represented simply as 1×1 real matrices) describe individual flavor-singlets and their antiparticle counterparts (e.g., u and $\bar{u}$ or e^- and e^+ etc.), evidently the 2×2 real matrices $\mathbf{F}$ provide a new description of [("up," "down")]- or {("down," "up")}-type flavor-doublets, i.e., [(u,d) and $(\bar{d},\bar{u})$] or {(e^-, ν_e) and $(\bar{\nu}_e, e^+)$}, respectively.

2.5.1 The "up"-"down" Flavor Dichotomy

As indicated in Table I, restricting the new internal-space to *two* dimensions already provides a possible "built-in" geometric "explanation" for the "up"-"down"-type flavor *dichotomy*. But, there are other dichotomies that can be given a simple geometric explanation as well.

2.5.2 The Matter-Antimatter Dichotomy

Given the definitions of electric-charge vectors [Eqs. (2.8) through (2.11) and Table I] and Eqs. (2.2) through (2.5), the quark (lepton) electric-charge labels q_1 and q_2 (q'_1 and q'_2) and the B (L) charge label, all change signs under the transformation: $q_1 \to -q_2$ and $q_2 \to -q_1$ ($q'_1 \to -q'_2$ and $q'_2 \to -q'_1$). As indicated in (1.17), such a transformation can be represented by a 2×2 matrix $-\boldsymbol{\sigma}_x$, which plays the role of *charge conjugation* $\mathbf{C}$. Limiting the following discussion to quarks, these results may be summarized by the equation

$$-\boldsymbol{\sigma}_x \{q_1, q_2\} = \{-q_2, -q_1\}, \tag{2.18}$$

and using the metric given by (1.26), also by the equation

$$B = \{q_1, q_2\}^2 = -\{-q_2, -q_1\}^2. \tag{2.19}$$

Clearly, the matter-antimatter dichotomy (see Appendices A and B) can be interpreted *geometrically* as being the result of an internal 2-space *reflection*—of a flavor-doublet as represented by an electric-charge vector $\mathbf{Q}_q = \{q_1, q_2\}$—induced by $-\boldsymbol{\sigma}_x$, about the line $x_2 = -x_1$.

In terms of flavor eigenstates

$$-\boldsymbol{\sigma}_x(u_{L,R}, d_{L,R}) = (\bar{d}_{R,L}, \bar{u}_{R,L}).$$

And, dropping the L and R (handedness) subscripts,

$$-\boldsymbol{\sigma}_x(u, d) = (\bar{d}, \bar{u}). \tag{2.20}$$

Equation (2.20) is strictly a *symbolic* transformation since $-\boldsymbol{\sigma}_x$ (unlike C) does *not* act directly on flavor eigenstates. Instead, it acts on electric-charge vectors and other real-vectors representing both individual *flavors* and *doublets* of flavor eigenstates. Similar statements apply to the leptons.

2.5.3 The Quark-Lepton "Dichotomy"

Although the *magnitudes* of the electric charges of quarks and leptons differ (as indicated by primes on the lepton charges), these particles can be interpreted *geometrically* as being *inverted* relative to each other. First, notice that the nonzero (primed and unprimed) charge-labels of Table I for matter and antimatter (bearing the subscript 1) are generally opposite in sign (e.g., compare the u and $\bar{u}$ charges of Table I with the e^- and e^+ charges, respectively). Second, notice that the coordinate-axis directions along which the (u, d) electric charges are measured, namely, the respective quadrant IV directions $(X_1, -X_2)$, are *inverted* relative to the coordinate-axis directions along which the (e^-, ν_e) electric charges are measured, namely, the respective quadrant II directions $(-X_1, +X_2)$. Thus, it could be argued from a *geometric* perspective, that the quark-lepton dichotomy (see Appendix A) owes its existence, at least in part, to an internal 2-space *inversion* $\mathbf{I} = -\mathbf{I}_2$ ($\mathbf{I}_2$ is the 2×2 identity matrix) of the electric-charge vectors, and other vectors, representing a quark or lepton flavor-doublet. Third, notice that the expressions for baryon number B can be transformed into the expressions for lepton number L [see Eqs. (2.2) through (2.5)] by a transformation that is the "product" of an inversion $\mathbf{I} = -\mathbf{I}_2$, and a hypothetical transformation $(v \rightarrow v')$ involving the dimensionless-parameters v and v' first introduced in (1.21). The latter transformation changes the *magnitudes* of the electric charges.

2.6. OTHER QUANTUM NUMBERS AND FAMILIES

The overall "product" transformation can be represented by the pair of coordinate replacements: $q_1 \to +q'_1, q_2 \to +q'_2$ for matter and $-q_1 \to -q'_1, -q_2 \to -q'_2$ for antimatter. Evidently, these replacements convert B to L, and reversing the indicated arrows, L to B (see Table I).

2.6 Other Quantum Numbers and Families in the 2-Space

As described in Sec. 1.4.1, the 2-vectors $\mathbf{Q}_q = \{q_1, q_2\}$ and $\mathbf{Q}_\ell = \{q'_1, q'_2\}$ represent quark and lepton flavor-doublets, respectively, and can be "resolved" into pairs of linearly-independent "basis" vectors $\mathbf{U}$ and $\mathbf{V}$ in more than one way. These other ways of resolving $\mathbf{Q}_q$ or $\mathbf{Q}_\ell$ provide an opportunity to introduce new flavor, or mass, eigenstates by introducing global flavor-defining charge labels *unrelated* to those associated with the first family.

2.6.1 Global Flavor-Defining Charge Labels

As indicated in (1.32), any 2-vector such as $\mathbf{Q}_q$ or $\mathbf{Q}_\ell$ can be expressed as

$$\mathbf{Q}_q \text{ or } \mathbf{Q}_\ell = \mathbf{U} + \mathbf{V}. \tag{2.21}$$

Now, just as the 2-vectors $\mathbf{Q}_q$ and $\mathbf{Q}_\ell$ represent individual flavors and flavor doublets (see Table I), the associated 2-vectors $\mathbf{U}$ and $\mathbf{V}$ also represent the *same* flavors and flavor doublets (see Sec. 1.4.1). Thus, the non-Euclidean *triad* of charge-vectors ($\mathbf{Q}_q$ or $\mathbf{Q}_\ell$, $\mathbf{U}$, $\mathbf{V}$) provides a representation of flavors and flavor doublets, which leads to correspondences such as

$$(u, d) \leftrightarrow \mathbf{Q}_q, \mathbf{U}, \mathbf{V}$$

or

$$(e^-, \nu_e) \leftrightarrow \mathbf{Q}_\ell, \mathbf{U}, \mathbf{V},$$

where the pair of linearly-independent charge-vectors $\mathbf{U}$ and $\mathbf{V}$ are *different* for quarks and leptons.

In general, the 2-vectors $\mathbf{U}$ and $\mathbf{V}$ serve to define new global charge-like scalar quantities [see Eq. (1.33)] associated with the baryon number B and the lepton number L since [see Eqs. (2.12) and (2.21)]

$$B = \mathbf{Q}_q^2 \text{ or } (\mathbf{Q}_q^c)^2 = \mathbf{U}^2 + 2\mathbf{U} \bullet \mathbf{V} + \mathbf{V}^2 \tag{2.22}$$

and

$$B' = L = \mathbf{Q}_\ell^2 \text{ or } (\mathbf{Q}_\ell^c)^2 = \mathbf{U}^2 + 2\mathbf{U} \bullet \mathbf{V} + \mathbf{V}^2, \tag{2.23}$$

where $\mathbf{U} \bullet \mathbf{U} = \mathbf{U}^2$ and $\mathbf{V} \bullet \mathbf{V} = \mathbf{V}^2$. Like B or L, each of the scalar-products appearing on the right-hand side of Eqs. (2.22) and (2.23) should be charge-like *quantum observables* that are "carried" by both flavors in a flavor doublet. Hence, the quantum-mechanical operators representing these observables should *commute* with the quantum-mechanical operators representing the quark or lepton electric-charges. Moreover, all of these charges should be "good" quantum numbers (constants of the motion) for "isolated" fermions and certain interactions.

Insofar as quarks carrying B are concerned, the expression for B in (2.22) is reminiscent of the expression for the *global hypercharge* $Y = B + S$ or *baryon number* $B = Y - S$ [see Eq. (1.31)] of $SU(3)$-flavor.[26,27] For example, Eqs. (2.22) and (2.23) will define global "generalized-hypercharges" for both quarks and leptons, respectively.[29] For strange quarks we will be able to show that strangeness $S = -\mathbf{U}^2$, hypercharge $Y = 2\mathbf{U} \bullet \mathbf{V}$ and $\mathbf{V}^2 = 0$.

2.6.2 Flavor Eigenstates

Using the results of Sec. 2.6.1, generalized quark-or-lepton flavor, or mass, eigenstates can be defined via ket-vectors similar in form to (see also Sec. 1.4.1)

$$|(q_1 \text{ or } q_2, B = \mathbf{Q}_q^2) \text{ or } [-q_1 \text{ or } -q_2, B = (\mathbf{Q}_q^c)^2], \mathbf{U}^2, 2\mathbf{U} \bullet \mathbf{V}, \mathbf{V}^2\rangle \tag{2.24}$$

or

$$|(q_1' \text{ or } q_2', L = \mathbf{Q}_\ell^2) \text{ or } [-q_1' \text{ or } -q_2', L = (\mathbf{Q}_\ell^c)^2], \mathbf{U}^2, 2\mathbf{U} \bullet \mathbf{V}, \mathbf{V}^2\rangle, \tag{2.25}$$

respectively, where $\mathbf{U}$ and $\mathbf{V}$ are appropriate to a particular flavor- or antiflavor-doublet. For example, it will be shown that the strange quark can be represented by a ket-vector similar to $|q_2, B, \mathbf{U}^2, 2\mathbf{U} \bullet \mathbf{V}, \mathbf{V}^2\rangle$, where $q_2 = -1/3$, $\mathbf{U} \bullet \mathbf{V} = Y/2 = (B+S)/2 = -1/3$, $B = \mathbf{Q}_q^2 = +1/3$, $S = -\mathbf{U}^2 = -1$ and $\mathbf{V}^2 = 0$. Similar ket-vectors will be used to represent the first family of quarks and leptons.

2.6.3 Effective "Quantization"

If the foregoing description of fundamental fermions is to agree with experiment, there must be some way to limit the number of ways $\mathbf{Q}_q$ or $\mathbf{Q}_\ell$ can be *resolved*. That is, there must be some way to effectively "quantize" the problem (see also Sec. 1.4.3) and, thereby, limit the number of physically acceptable charge-vector triads ($\mathbf{Q}_q$ or $\mathbf{Q}_\ell, \mathbf{U}, \mathbf{V}$). These charge-vectors will be limited if there is some way to "quantize" their *components*.

By definition, the quantized, C-reversing, global charges carried by fundamental fermions can have the value zero (the exceptions being B, L and f, which are *never* zero for fermions) or a number of different positive or negative nonzero-values, namely, $0, \pm a, \pm b, \pm c, \ldots$, where $a, b, c, \ldots$, are dimensionless real-numbers to be determined by "quantum constraints." These constraints fall into one of three categories:

1) Simple physical constraints, e.g., it must be possible to *distinguish* flavors using these charges *alone*.

2) "Standard-model" constraints, e.g., the fundamental unit of electric charge is given by the "quantum condition" $q_1 - q_2 = -q'_1 - q'_2 = 1$, and the individual electric-charges are constrained by the "quantum condition" $v = \ln M_c$, where $M_c = 3$ for quarks and $M_c = 1$ for leptons.

3) Algebraic and geometric constraints dictated by the non-Euclidean geometry, e.g., the charges $\mathbf{Q}^2$, $\mathbf{U}^2$, $\mathbf{V}^2$, $\mathbf{U} \bullet \mathbf{V}$ transform, in the 2-space, like non-Euclidean scalar-products.

In Chapters 4 and 5, it will be shown that the number of flavors, flavor-doublets and families described by the present scheme is *finite* in number. In particular, the foregoing algebraic, geometric and physical constraints will limit the number of $\mathbf{U}$-vectors to six. This limits the number of flavor doublets to six and the number of families to three. Then, the $\mathbf{Q}$-vectors and the vectors $\mathbf{V} = \mathbf{Q} - \mathbf{U}$ will be *deduced* by solving eigenvalue equations similar to (2.15) and (2.16) for $\mathbf{Q}$, when $\mathbf{F}$ is given by the upper signs in (1.21). Note that $\mathbf{Q}$-vectors, and their component electric-charges, were previously *assumed* or taken from experiment (see Table I).

Before taking up the question of effective "quantization" in more detail, it will be helpful to have a better understanding of the 2-space geometry and topology. This is accomplished in the next chapter by representing the 2-space on a Euclidean plane ($E2$) using the labeling conventions of Table I.

To discover the theory of evolution, the animal kingdom had first to be classified. To understand the atom, the chemical elements had first to be organized in the periodic table. To understand the 1/r potential of gravitation, Newton had to follow Kepler and the patterns he registered and systematized.

—Y. Ne'eman

3. Graphical Representation of the New 2-Space

The 2-space introduced in Chapters 1 and 2 can be represented graphically on the Euclidean plane ($E2$) using the four flavor-labels and electric charges associated with first-family fundamental fermions (see Table I). When this method of representation is chosen, the first- and third-quadrants of the 2-space (defined by the X_1 and X_2 coordinate axes) are *unphysical*, while the second- and third-quadrants are *physical*. Moreover, the physical regions of the 2-space are found to consist of four *topologically-distinct* subregions or *octants*. Thus, the *measurement directions* (see Sec. 2.6.3 and Table I), which *define* the physical regions of the 2-space, are "preferred" directions. Hereinafter, these directions will be referred to as *flavor directions*.

3.1 Flavor and Antiflavor Directions

As indicated in Table I, the electric charges of the four first-family flavors (u, d, e^-, ν_e) are "measured" along four coordinate-axis directions, namely, $(+X_1, -X_2, -X_1, +X_2)$, respectively. Because of the correspondence between coordinate-axis directions in the 2-space and flavors, these measurement directions will also be called *flavor directions* and are labeled accordingly (see Table I and Fig. 3.1a). Similarly, the set of four first-family antiflavors $(\bar{d}, \bar{u}, \bar{\nu}_e, e^+)$ correspond, respectively, to the *same* four coordinate-axis directions $(+X_1, -X_2, -X_1, +X_2)$. These directions are now referred to as *antiflavor directions* (see Table I and Fig. 3.1b).

Evidently, some of the flavor- and antiflavor-directions are the *same* (see Fig. 3.1c). But, this does *not* mean that the corresponding particles or flavor eigenstates are the same, only that their electric charges and other charges (e.g., components of **U** and **V**), are "*measured*" along the *same* direction.

By inspection, the antiflavor directions of Fig. 3.1b are related to the corresponding flavor-directions of Fig. 3.1a by the matter-antimatter transformation $-\boldsymbol{\sigma}_x$ [see Table I and Eq. (1.17)], which induces a "reflection" of *any* 2-vector (e.g., vectors from the origin pointing along any direction) about the diagonal line $x_2 = -x_1$. Note that this line is a one-dimensional *invariant vector-space*[30] defined by the infinity of eigenvectors (eigenvalue +1) of the matrix $-\boldsymbol{\sigma}_x$ having the form $\{k, -k\}$, where k is any real number. Thus, $-\boldsymbol{\sigma}_x$ *symbolically* transforms quark and lepton flavor-symbols (flavor directions) into antiquark or antilepton flavor-symbols (antiflavor directions), respectively, and vice versa [see also Eq. (2.20)].

3.2 Unphysical Regions of the 2-Space

Because of the particular choices for electric-charge labels or *subscripts* (i.e., 1 or 2) associated with matter (see Table I), certain regions of the internal 2-space illustrated in Fig. 3.1 are *not* physically meaningful. In general, charge-vectors $\{x_1, x_2\}$ can only represent individual quark (lepton) flavors and/or flavor doublets when *both* the x_1 and x_2 internal coordinates correspond to charge labels (e.g., electric-charge labels or *other* internal charges) carried by quarks (leptons). When x_1 corresponds to a quark charge-label and x_2 corresponds to a lepton charge-label, or vice versa, the vectors $\{x_1, x_2\}$ cannot be used to represent individual quark or lepton flavors and/or flavor doublets. For this reason, among others to be cited in the next paragraph, such vectors are considered to be *unphysical.* Therefore, vectors $\{x_1, x_2\}$ in quadrant II ($x_1 < 0, x_2 > 0$) of Fig. 3.1 represent lepton or antilepton flavor-doublets, while vectors $\{x_1, x_2\}$ in quadrant IV ($x_1 > 0, x_2 < 0$) represent quark or antiquark flavor-doublets, whereas vectors $\{x_1, x_2\}$ in the shaded regions, i.e., quadrants I ($x_1 > 0, x_2 > 0$) or III ($x_1 < 0, x_2 < 0$), are unphysical and do not represent flavor doublets.

Another way to see that quadrants I and III are unphysical, is to notice that a "reflection" $+\boldsymbol{\sigma}_x$ of the two internal coordinate-axes about the line $x_2 = x_1$ (this is the analog of the "reflection" $-\boldsymbol{\sigma}_x$ about $x_2 = -x_1$) does *not* change the signs of the components of vectors originating at the origin whose "tips" lie in quadrants I or III. Hence, *neither* of these components act like C-reversing charges and, therefore, these vectors cannot be charge vectors. Moreover, a scalar-product such as $\{x_1, x_2\}^2$ cannot represent B or L (even though it *does* change signs under $+\boldsymbol{\sigma}_x$), since x_1 and x_2 involve both quark *and* lepton charges. These results mean that *transformations analogous to*

36   3. GRAPHICAL REPRESENTATION

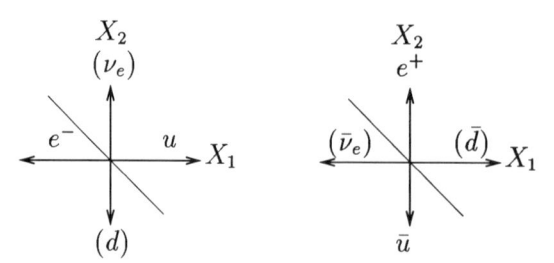

a) Four flavor directions   b) Four antiflavor directions

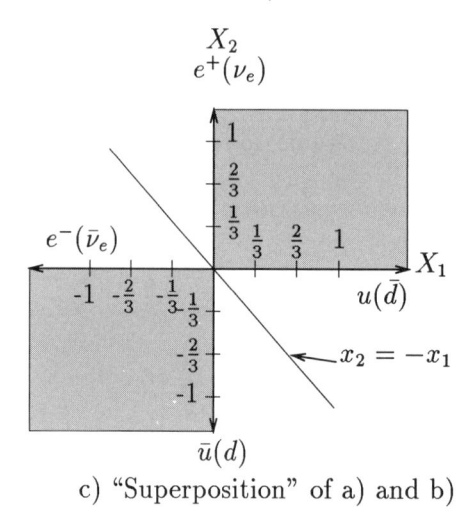

c) "Superposition" of a) and b)

Figure 3.1. An abstract, internal non-Euclidean 2-space is represented on the Euclidean plane (E2). In a) and b) the 2-space is *defined* by the electric charges carried by the four first-family flavor (and antiflavor) eigenstates of Table I. An "up" ("down")-type "flavor direction" is the direction along which "up" ("down")-type electric-charges, including zero and nonzero values, are *measured*. In c) the positive abscissa $x_1 \geq 0$ is labeled by the "up" ("antidown")-type quark (antiquark) state $u(\bar{d})$, which carries positive fractional-charge $2/3(1/3)$, while the negative abscissa $x_1 \leq 0$ is labeled by the "down" ("antiup")-type lepton (antilepton) state $e^-(\bar{\nu}_e)$, which carries negative (neutral) integral-charge $-1(0)$. The positive ordinate $x_2 \geq 0$ is labeled by the "antidown" ("up")-type antilepton (lepton) state $e^+(\nu_e)$, which carries positive (neutral) integral-charge $+1(0)$, while the negative ordinate $x_2 \leq 0$ is labeled by the "antiup"("down")-type antiquark (quark) state $\bar{u}(d)$, which carries negative fractional-charge $-2/3(-1/3)$. "Antiup" or "antidown" states may also be consistently referred to as "down" or "up" states, respectively.

3.3 Other Geometric and Topological Properties of the 2-Space

By dropping the flavor labels on the X_1 and X_2 axes of Fig. 3.1, and displaying the fixed (constant) charge-vectors $\mathbf{Q}_q, \mathbf{Q}_q^c, \mathbf{Q}_\ell$ and $\mathbf{Q}_\ell^c$, written as functions of v and f or v' and f [see Eq. (1.21)], additional geometric (and topological) aspects of the physical regions of the internal 2-space can be illustrated as in Fig. 3.2. Quark or antiquark electric-charge vectors $[\mathbf{Q}_q$ or $\mathbf{Q}_q^c = \mathbf{Q}_q(v,f)]$ are designated, respectively, $\mathbf{Q}_q = \mathbf{Q}_q(v,1)$ or $\mathbf{Q}_q^c = \mathbf{Q}_q(v,-1)$, while lepton or antilepton electric-charge vectors $[\mathbf{Q}_\ell$ or $\mathbf{Q}_\ell^c = \mathbf{Q}_\ell(v',f)]$ are designated, respectively, $\mathbf{Q}_\ell = \mathbf{Q}_\ell(v',1)$ or $\mathbf{Q}_\ell^c = \mathbf{Q}_\ell(v',-1)$.

Because singly-charged weak intermediate-vector-bosons $W^\pm$ connect "up"- and "down"-type flavor eigenstates in a flavor doublet, "up"- and "down"-type electric-charges differ by the fundamental unit of electric charge[4] (e.g., the weak-color charge $G = \pm 1$). Therefore, the "tips" of the electric-charge vectors $\mathbf{Q}_q$ and $\mathbf{Q}_q^c$ must fall on the straight line (see Fig. 3.2)

$$q_1 - q_2 = 1, \tag{3.1}$$

while the "tips" of the electric-charge vectors $\mathbf{Q}_\ell$ and $\mathbf{Q}_\ell^c$ must fall on the straight line

$$q_1' - q_2' = -1. \tag{3.2}$$

The "quantum conditions" expressed by (3.1) and (3.2) are left invariant by $-\boldsymbol{\sigma}_x$ since this reflection reverses subscripts 1 and 2 and simultaneously changes the signs of the electric-charge labels [see Eqs. (1.18) and (1.19)].

The physical regions of the internal 2-space (quadrants II and IV) consist of four *topologically-distinct* subregions (octants). These octants are topologically-distinct because the electric-charge vectors $\mathbf{Q}_q$ or $\mathbf{Q}_\ell$ cannot be *continuously* transformed [by transformations associated with the metric g of Eq. (1.26)] into their antiparticle counterparts $\mathbf{Q}_q^c$ or $\mathbf{Q}_\ell^c$, respectively, nor can they be continuously transformed into $\mathbf{Q}_\ell$ or $\mathbf{Q}_q$, respectively.

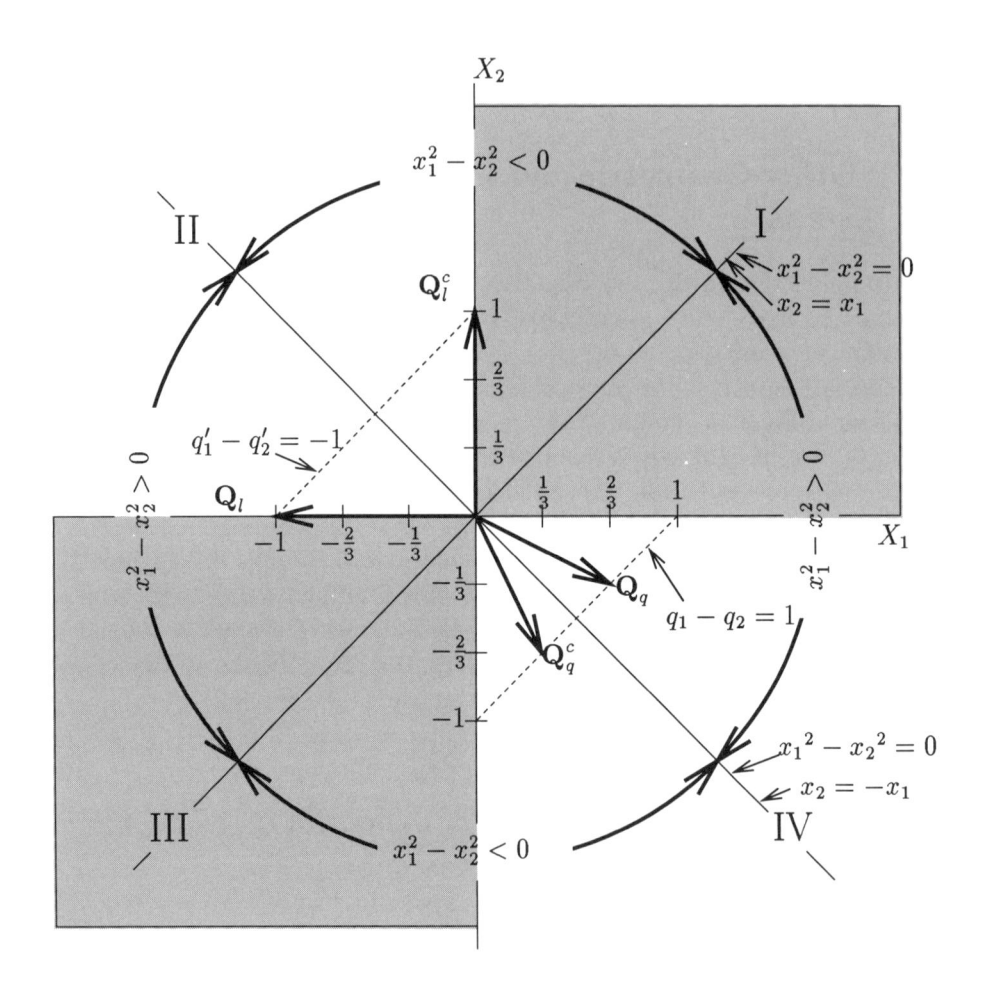

Figure 3.2. Additional geometrical and topological aspects of the two-dimensional non-Euclidean linear vector-space of Fig. 3.1 are illustrated. The one-dimensional invariant vector-spaces associated with the reflection matrices ($\pm\boldsymbol{\sigma}_z$) and ($\pm\boldsymbol{\sigma}_x$) define the coordinate axes (X_1, X_2) and the diagonal lines ($x_1 = \pm x_2$), respectively.[30] Given the non-Euclidean metric **g** of (1.26), these one-dimensional invariant vector-spaces naturally divide the plane into 8 topologically-distinct subregions or octants. See Sec. 3.3 in the main text for further details. Electric-charge vectors $\mathbf{Q}_q$, $\mathbf{Q}_q^c$, $\mathbf{Q}_l$ and $\mathbf{Q}_l^c$ are indicated.

There ought to be something very distinctive about the theory that describes the Universe. Why does this theory come to life while other theories exist only in the minds of their inventors?

—S. W. Hawking

4. Effective "Quantization" of the U-Vector Components and Three Families

In this and subsequent chapters I will attempt to "derive" families from the algebraic and geometric properties of the 2-space (see Figs. 3.1 and 3.2). In particular, if all physically acceptable **U**- and **V**-vectors (representing flavor doublets) can somehow be derived from the 2-space geometry, then the flavors, flavor-doublets and families (including antiparticles) that are consistent with this geometry, will have been derived.

Unfortunately, because the underlying dynamics responsible for families (Planck-level dynamics?) is largely unknown, no first-principles *quantization* of the components of **U** and **V** is possible at this time. But, because the same underlying dynamics is presumably responsible for the algebraic and geometric constraints associated with the new 2-space, these constraints could conceivably embody an effective "quantization" of the foregoing **U**- and **V**-vector components, thereby, leading to a determination of these vectors without ever having to deal with the underlying dynamics directly.

4.1 The Vectors U and V

According to the discussion in Sec. 1.4.1, it takes four scalar charge-labels (i.e., the quantum numbers u_1, u_2, v_1 and v_2 for quarks) to specify each of two flavor-eigenstates in any given flavor-doublet (e.g., $|u_1, v_1\rangle$ and $|u_2, v_2\rangle$) and to distinguish different flavor doublets. But, which of these charge-labels provides the distinction between "up"- and "down"-type flavor eigenstates within the *same* flavor doublet?

While two charge-label types, namely, u (u_1 or u_2) and v (v_1 or v_2) are available for this purpose, only one type (i.e., u or v) would suffice. Accordingly, I make the simplifying assumptions that $v_1 = v_2$ and $u_1 \neq u_2$ *for all possible flavor doublets.* That is, a "minimal" set of *three* (instead of four) independent *scalar charge-labels* (i.e., $u_1, u_2, v_1 = v_2$) are now assumed to specify each of two flavor-eigenstates in any given flavor-doublet, and to distinguish different flavor doublets. This reduction in the number of flavor-defining charges means that *the scalar charge-label u, and only this label, distinguishes "up"- and "down"-type flavors within the same flavor doublet.*

4.1.1 The Form of U and V

Like all *quantized* C-reversing "charges," u-charge should have a multiplicity of values including zero and nonzero (positive and negative) values. That is, *fundamental fermions will either be "neutral" or "charged" with respect to u-charge.*

The results of Sec. 1.4.1 show that for quarks: $q_1 - q_2 = u_1 - u_2$ when $v_1 = v_2$; and a similar expression applies to the leptons. Consistency with Eqs. (3.1) and (3.2) permits the linearly-independent vectors **U** and **V** to be redefined.

Using the symbols a, a', Y and $Y' \neq Y$,

$$\left. \begin{array}{rl} \mathbf{U} &= \{a, a-1\} \\ \\ \mathbf{V} &= \{Y/2, Y/2\} \end{array} \right\} \quad (4.1)$$

for quarks, and

$$\left. \begin{array}{rl} \mathbf{U} &= \{a', a'+1\} = \{-a, -a+1\} \\ \\ \mathbf{V} &= \{-Y'/2, -Y'/2\} \end{array} \right\} \quad (4.2)$$

for leptons, which are "inverted" relative to quarks (see Sec. 2.5.3 and Fig. 3.1a). Here, the "up" ("down")-type quark charge-label is $a(a-1)$, while the corresponding *inverted* "down" ("up")-type lepton charge-label is $a'(a'+1)$.

The 2-vector form (4.1) is compatible with the well-known Gell-Mann–Nishijima charge-relation of $SU(3)$-flavor,[26,27] wherein the electric charges of the first-family u and d quarks are given by (1.30). In (1.30) the third-component of global isospin T_3 is different for the u and d quarks, thereby allowing them to be distinguished. Similarly, the quantum numbers or u-type charges a and $a-1$ in (4.1) distinguish between u and d quarks, respectively. By contrast, the global hypercharge in $SU(3)$-flavor [see Eq. (1.31)], like Y

4.1. THE VECTORS U AND V

in (4.1), is the same for the u and d quarks.[26,27] Hence, Y in (4.1) provides no distinction between the u and d quarks.

The foregoing considerations mean that Y and Y' in Eqs. (4.1) and (4.2), respectively, play the roles of global generalized-hypercharge labels, while the "up" ("down")-type charge-labels a and $a'+1$ ($a-1$ and a') play the roles of special, global, generalized charge-labels analogous to global T_3, which serve to distinguish "up"- and "down"-type flavor-eigenstates within the same flavor doublet.

Given Eqs. (4.1) and (4.2) and the metric **g** in (1.26), we *always* have the charge

$$\mathbf{V}^2 = 0, \tag{4.3}$$

and the charges

$$\mathbf{U} \bullet \mathbf{V} = Y/2 \text{ or } Y'/2. \tag{4.4}$$

From (4.3) we see that *all fundamental fermions are neutral (i.e., they act like "singlets") with respect to the charge* $\mathbf{V}^2$.

Using Eqs. (4.1) and (4.2), we can simplify B and L, as originally expressed by Eqs. (2.22) and (2.23), respectively, to read

$$B = \mathbf{U}^2 + Y \tag{4.5}$$

and

$$B' = L = \mathbf{U}^2 + Y'. \tag{4.6}$$

The resultant Gell-Mann–Nishijima-like charge relations[26,27] for the electric-charge labels are [see Eqs. (4.1), (4.2) and Sec. 1.4.1]

$$q_1 = a + Y/2 \tag{4.7}$$

and

$$q_2 = (a-1) + Y/2 \tag{4.8}$$

for quarks, and

$$q'_1 = -a - Y'/2 \tag{4.9}$$

and

$$q'_2 = (-a+1) - Y'/2 \tag{4.10}$$

for the *corresponding* "inverted" leptons in the *same* family.

4.2 The Six Physically Acceptable U-Vectors

In the following I will show that the physically acceptable **U**-vectors, necessarily, reside in the one-dimensional *invariant vector-spaces*[30] associated with the matrices $-\boldsymbol{\sigma}_x$ or $\pm\boldsymbol{\sigma}_z$ (see Fig. 3.2). The first-family **U**-vectors are shown to be eigenvectors of $-\boldsymbol{\sigma}_x$, while the second- and third-family **U**-vectors are shown to be eigenvectors of $+\boldsymbol{\sigma}_z$ and $-\boldsymbol{\sigma}_z$, respectively, (all eigenvalues +1).

Like all *quantized* **C**-reversing charges, the "charge" $\mathbf{U}^2$ should have a multiplicity of values, including zero and nonzero (positive and negative) values. That is, *both members of a flavor doublet will either be "neutral" or "charged" (carry the same nonzero charge) with respect to* $\mathbf{U}^2$*-charge.*

In the absence of constraints on its value other than Eqs. (4.1), (4.2) and (1.26), the charge

$$\mathbf{U}^2 = 2a - 1 \text{ or } -2a' - 1, \tag{4.11}$$

can have the value *zero* or it can have a *continuum* of positive or negative *nonzero* values.

4.2.1 The First Family

Clearly, when $\mathbf{U}^2$ is zero in (4.11), a and a' are effectively "*quantized*," i.e., $a = 1/2$ or $a' = -a = -1/2$ only. Thus, when $\mathbf{U}^2 = 0$ for quarks, the only way to distinguish "up"-type flavors from "down"-type flavors, respectively, is for the "up"-type charge-label a to be different in sign, but equal in magnitude, to the "down"-type charge-label $a - 1$. Therefore, *there are only two physically acceptable* **U**-*vectors associated with the "quantized"* $\mathbf{U}^2$*-value zero*, namely,

$$\mathbf{U} = \{1/2, -1/2\} \text{ and } \{-1/2, 1/2\}. \tag{4.12}$$

These vectors are the additive *inverses* of one another (i.e., one describes quarks and the other describes leptons) and both vectors are eigenvectors (eigenvalue +1) of the matter-antimatter reflection $-\boldsymbol{\sigma}_x$ [see Eq. (1.17)]. Both of these eigenvectors lie "half-way" between the X_1 and X_2 internal coordinate-axes indicated in Figs. 3.1, 3.2 and 4.1. That is, these vectors are directed from the origin along the line $x_2 = -x_1$. We will later identify the third-component of global-isospin associated with the u (d) quark [as in $SU(3)$-flavor[26,27]] with the component $a(a-1)$, and by generalizing to lepton

4.2. THE SIX PHYSICALLY ACCEPTABLE U-VECTORS 43

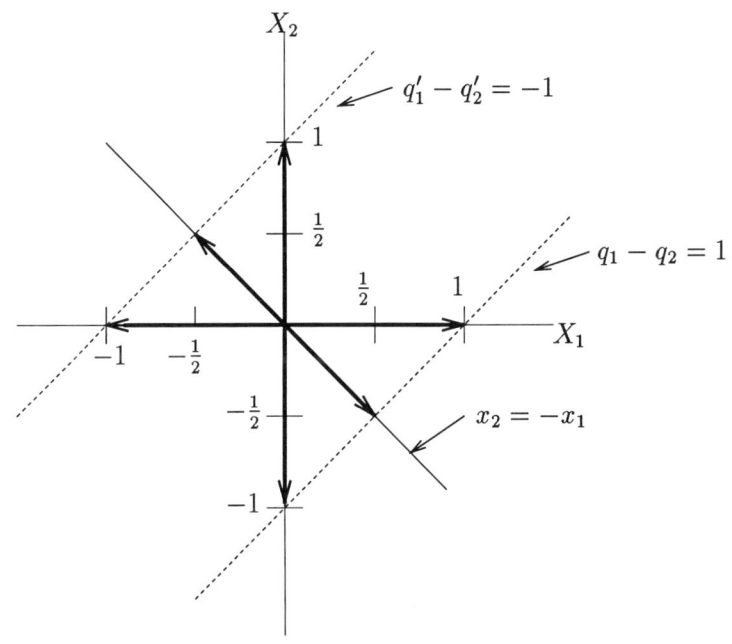

Figure 4.1. The six physically acceptable **U**-vectors (indicated by long bold arrows) lie in the one-dimensional *invariant vector-spaces* associated with the matrices $-\boldsymbol{\sigma}_x$ or $\pm\boldsymbol{\sigma}_z$. Only these six **U**-vectors [see Eqs. (4.12) and (4.13)], from an otherwise infinite set of **U**-vectors in the 2-space, permit "up"- and "down"-type quark or lepton flavors to be distinguished. These six vectors are either eigenvectors of the reflection $-\boldsymbol{\sigma}_x$ about the line $x_2 = -x_1$, or they are eigenvectors of the reflection $-\boldsymbol{\sigma}_z$ $(+\boldsymbol{\sigma}_z)$ about the line $x_1 = 0$ $(x_2 = 0)$. The "tips" of all **U**-vectors are constrained by Eqs. (3.1), (3.2), (4.1) and (4.2), to fall on one of the two sloping dotted lines indicated. Note that all eigenvalues associated with the eigenvectors referred to here equal +1.

global "isospin," the component a' $(a'+1)$ will be identified with the e^- (ν_e) lepton.[29] Thus, the "quantum condition" $\mathbf{U}^2 = 0$ leads us to the first family of fundamental fermions. And, therefore, *every fundamental fermion (quark or lepton) in the first family carries the same numerical charge* $\mathbf{U}^2 = 0$. Evidently, when the charge $\mathbf{U}^2$ is nonzero, we have constraints on additional families beyond the first family.

4.2.2 The Second and Third Families

Consider the case when the charge $\mathbf{U}^2 \neq 0$ and one member of a flavor doublet is *neutral* with respect to u-charge (see Sec. 4.1). Since $\mathbf{U}^2 \neq 0$, it immediately follows that the other member of the same flavor-doublet must be *charged* with respect to u-charge so that "up"- and "down"-type flavors in the flavor doublet can be distinguished.

From these constraints, and Eqs. (4.1), (4.2) and (4.11), it follows that *the charges a and a' assume the effectively "quantized" values $a = +1$ or 0 and $a' = -1$ or 0 only*. Thus, instead of an infinite number of possible U-vectors associated with nonzero values of $\mathbf{U}^2$, there are actually only four physically-acceptable U-vectors. These U-vectors are

$$\mathbf{U} = \{1,0\}, \{0,-1\} \text{ and } \{-1,0\}, \{0,1\}. \tag{4.13}$$

Each of these U-vectors is an eigenvector of $+\boldsymbol{\sigma}_z$ or $-\boldsymbol{\sigma}_z$ with eigenvalue $+1$, subject to the constraints (3.1) and (3.2), and consequently the "quantized" $\mathbf{U}^2$-values are limited to ± 1. That is, *every fundamental-fermion (quark or lepton) in what we will later call the second or third families, carries the same numerical charge* $\mathbf{U}^2 = +1$ or $\mathbf{U}^2 = -1$, *respectively*. Combining Eqs. (4.12) and (4.13) we see that, in general, the "quantized" $\mathbf{U}^2$-values are limited to $0, \pm 1$, while the "quantized" u-values are limited to $0, \pm \frac{1}{2}, \pm 1$. Thus, we find that *all fundamental fermions act like "triplets" (or possibly a "singlet" and a "doublet") with respect to the charge $\mathbf{U}^2$, and this charge has the same numerical value for all members of a given family*.

Even though the charge $\mathbf{U}^2$ is *numerically* the same for quarks and leptons in the same family, we will show that it has a different *qualitative* significance for quarks and leptons. That is, because the charge $\mathbf{U}^2$ is taken to be *separately* conserved for quarks and leptons, it must be *qualitatively* different for these particles.

Does equation (4.13) exhaust the possibilities or are there other U-vectors whose component u-charges are *both* nonzero, but $\mathbf{U}^2 \neq 0$? If so, both flavors

4.2. THE SIX PHYSICALLY ACCEPTABLE U-VECTORS 45

in a flavor doublet described by $\mathbf{U} = \{u_1, u_2\}$ would be "charged" with respect to u. But, this would be inconsistent (except when $\mathbf{U}^2 = 0$) with the proposal (see Sec. 4.1) that u-charge *alone* is responsible for distinguishing "up"- and "down"-type flavors. For example, since $u_1 - u_2 = \pm 1$, if both $u_1 \neq 0$ and $u_2 \neq 0$, we would have complicated **U**-vectors such as $\{6, 5\}$ or $\{-1, -2\}$, where *both* flavors in a flavor-doublet carry different amounts of the *same* kind of u-charge. Under these circumstances, no clear distinction between "up"- and "down"-type flavors could be made. Because of such inconsistencies, vectors of this type can have no physical significance. Hence, *Eqs. (4.12) and (4.13) define the only physically acceptable* **U**-*vectors in this scheme.*

As noted earlier, each of the vectors in (4.13) happens to be an eigenvector of a transformation (2-space reflection) represented by the Pauli matrix $\boldsymbol{\sigma}_z$, which is also the matrix representing the metric [see Eq. (1.26)]. Accordingly, each vector in (4.13) is directed from the origin along one of the two internal coordinate-axes, namely, the X_1 or X_2 axes, as indicated in Figs. 3.1 and 4.1.[30]

The first pair of vectors indicated on the left in (4.13) are, respectively, the additive inverses of the second pair on the right in (4.13). And, because of the matter-antimatter reflection $-\boldsymbol{\sigma}_x$, each of the six **U**-vectors indicated in Eqs. (4.12), (4.13) and Fig. 4.1, should apply to both *matter* and *antimatter* flavor-doublets (e.g., see Fig. 3.1a, 3.1b and 3.1c). Finally, because of the action of the inversion $\mathbf{I} = -\mathbf{I}_2$, each of the three **U**-vectors in quadrant IV describing a quark (antiquark) flavor-doublet[30,32] has an inverted counterpart describing a lepton (antilepton) flavor-doublet in quadrant II and vice versa (see Fig. 3.1). Hence, *only the* six **U**-*vectors in Eqs. (4.12) and (4.13) represent flavor doublets.*

The foregoing very general considerations mean that there can only be *three* quark (antiquark) flavor-doublets and *three* corresponding "inverted" lepton (antilepton) flavor-doublets. Hence, we have shown that, with the possible exception of families associated with very massive neutrinos (see Sec. 1.0, p. 1), *only twelve distinguishable fundamental-fermion (antifermion) flavors and three quark-lepton families (antifamilies) can be self-consistently represented in the internal non-Euclidean 2-space.* This, then, is the central *ex post facto* "prediction" of the new organizing principle stated in Sec. 1.2.6. The exact identification of which families a particular particle or flavor "resides in" will be taken up in Chapter 6, where the four **U**-vectors in (4.13) will be identified with two additional quark-lepton families (and their antiparticles). That is, the components of these vectors will be identified with flavor-defining quantum numbers or charge labels such as *charm* and *beauty*.

4.2.3 Strangeness and Truth

Now that the 2-space algebra and geometry has provided a partial "explanation" for two additional families beyond the first family, we can legitimately anticipate the incorporation of the known s and t quarks. To do this, we now define the generalized *strangeness* and *truth* charge labels S and t, respectively, by the equations

$$S + t = -\mathbf{U}^2 \qquad (4.14)$$

for quarks, and

$$S' + t' = -\mathbf{U}^2 \qquad (4.15)$$

for leptons, where $\mathbf{U}$ represents a quark flavor-doublet and $-\mathbf{U}$ represents the *corresponding* lepton flavor-doublet in the *same* family.

These quantum numbers are said to be "generalized" because leptons (primed quantities), in addition to quarks (unprimed quantities), are both being incorporated.[29] Here, the numerical charges $S = S' \neq 0$ and $t = t' = 0$ are taken to describe the *second* family, while the numerical charges $t = t' \neq 0$ and $S = S' = 0$ are taken to describe the *third* family (see Fig. 4.2). And, even though S and S' (t and t') are *numerically* the same for all fundamental fermions in the second (third) family, they should be thought of as being *qualitatively* different charges since they are taken to be *separately* conserved. This supposition is supported by the fact that S and t are associated with strongly-interacting quarks carrying baryon-number B, whereas S' and t', respectively, are associated with electroweakly-interacting leptons carrying lepton-number L.

That the S and t (S' and t') charges must be *qualitatively* different and easily distinguished, can be established by noting that numerically identical S and t (S' and t') are always associated with values of $B(L)$ having *opposite* signs. That is, if S and t (S' and t') are numerically the same, then one charge is always associated with *matter* while the other charge is always associated with *antimatter* (see Fig. 4.2). That the S and t charge labels in (4.14) are similar to the conventional strangeness [of $SU(3)$-flavor][26,27] and truth quantum numbers, will be demonstrated later when specific numerical values of S and t are deduced.

4.2. THE SIX PHYSICALLY ACCEPTABLE U-VECTORS 47

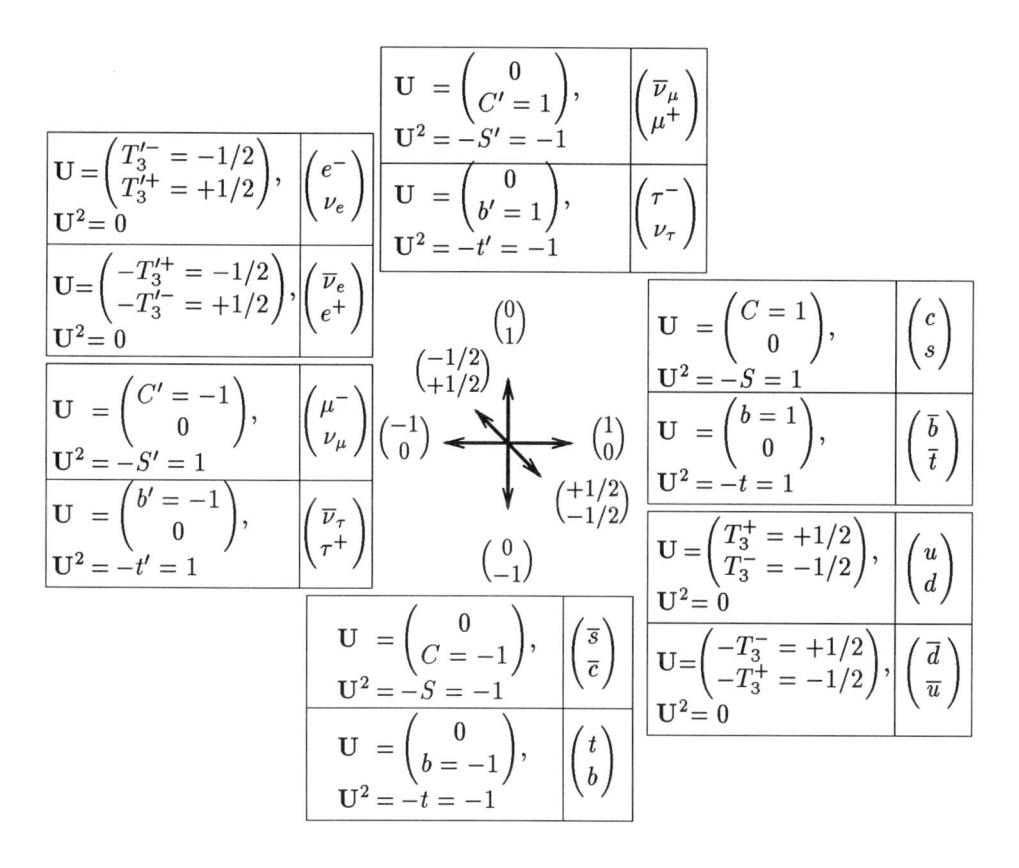

$\mathbf{U} = \begin{pmatrix} T_3'^- = -1/2 \\ T_3'^+ = +1/2 \end{pmatrix}$, $\mathbf{U}^2 = 0$	$\begin{pmatrix} e^- \\ \nu_e \end{pmatrix}$	
$\mathbf{U} = \begin{pmatrix} -T_3'^+ = -1/2 \\ -T_3'^- = +1/2 \end{pmatrix}$, $\mathbf{U}^2 = 0$	$\begin{pmatrix} \bar{\nu}_e \\ e^+ \end{pmatrix}$	
$\mathbf{U} = \begin{pmatrix} C' = -1 \\ 0 \end{pmatrix}$, $\mathbf{U}^2 = -S' = 1$	$\begin{pmatrix} \mu^- \\ \nu_\mu \end{pmatrix}$	
$\mathbf{U} = \begin{pmatrix} b' = -1 \\ 0 \end{pmatrix}$, $\mathbf{U}^2 = -t' = 1$	$\begin{pmatrix} \bar{\nu}_\tau \\ \tau^+ \end{pmatrix}$	

Top center:
$\mathbf{U} = \begin{pmatrix} 0 \\ C' = 1 \end{pmatrix}$, $\mathbf{U}^2 = -S' = -1$ — $\begin{pmatrix} \bar{\nu}_\mu \\ \mu^+ \end{pmatrix}$

$\mathbf{U} = \begin{pmatrix} 0 \\ b' = 1 \end{pmatrix}$, $\mathbf{U}^2 = -t' = -1$ — $\begin{pmatrix} \tau^- \\ \nu_\tau \end{pmatrix}$

Center compass: $\begin{pmatrix} 0 \\ 1 \end{pmatrix}$, $\begin{pmatrix} -1/2 \\ +1/2 \end{pmatrix}$, $\begin{pmatrix} -1 \\ 0 \end{pmatrix}$, $\begin{pmatrix} 1 \\ 0 \end{pmatrix}$, $\begin{pmatrix} +1/2 \\ -1/2 \end{pmatrix}$, $\begin{pmatrix} 0 \\ -1 \end{pmatrix}$

Right side:
$\mathbf{U} = \begin{pmatrix} C = 1 \\ 0 \end{pmatrix}$, $\mathbf{U}^2 = -S = 1$ — $\begin{pmatrix} c \\ s \end{pmatrix}$

$\mathbf{U} = \begin{pmatrix} b = 1 \\ 0 \end{pmatrix}$, $\mathbf{U}^2 = -t = 1$ — $\begin{pmatrix} \bar{b} \\ \bar{t} \end{pmatrix}$

$\mathbf{U} = \begin{pmatrix} T_3^+ = +1/2 \\ T_3^- = -1/2 \end{pmatrix}$, $\mathbf{U}^2 = 0$ — $\begin{pmatrix} u \\ d \end{pmatrix}$

$\mathbf{U} = \begin{pmatrix} -T_3^- = +1/2 \\ -T_3^+ = -1/2 \end{pmatrix}$, $\mathbf{U}^2 = 0$ — $\begin{pmatrix} \bar{d} \\ \bar{u} \end{pmatrix}$

Bottom:
$\mathbf{U} = \begin{pmatrix} 0 \\ C = -1 \end{pmatrix}$, $\mathbf{U}^2 = -S = -1$ — $\begin{pmatrix} \bar{s} \\ \bar{c} \end{pmatrix}$

$\mathbf{U} = \begin{pmatrix} 0 \\ b = -1 \end{pmatrix}$, $\mathbf{U}^2 = -t = -1$ — $\begin{pmatrix} t \\ b \end{pmatrix}$

Figure 4.2. Anticipating the introduction of two additional families, this figure illustrates **U**-vectors, their components (u-charge), and the scalar-product $\mathbf{U}^2$ for associated flavor doublets. Scalar components of **U**, the scalar $\mathbf{U}^2$, and the baryon number B, or lepton number L, are collectively taken to be conserved "charges" that distinguish flavor doublets in different families (see Sec. 4.2.3). Six fundamental-fermion flavor-doublets (3 quark and 3 lepton doublets), together with their antiparticle counterparts are, thereby, *uniquely* defined. The names (symbols) of the various flavor-eigenstates in this figure are the same as those of the experimentally known particles. These choices will be further justified later when other conserved "charges" (e.g., C, C', b, b'), and other geometric-associations, have been more fully explained.

Combining Eqs. (4.5) and (4.6) with Eqs. (4.14) and (4.15), respectively, we are led to define the global generalized-hypercharge

$$Y = B + S + t \tag{4.16}$$

for quarks, and

$$Y' = B' + S' + t' \tag{4.17}$$

for leptons.

Since B ($B' = L$) is the same for all quark (lepton) flavor-doublets, it is clear that $S + t = -\mathbf{U}^2$ and $\mathbf{U}$ ($S' + t' = -\mathbf{U}^2$ and $\mathbf{U}$) provide the distinction between quark (lepton) flavor-doublets in different families. As illustrated in Fig. 4.2, the components C and b (C' and b') of $\mathbf{U}$, like S and t (S' and t'), also serve as *qualitatively* different charges, which we take to be *separately* conserved in all but some weak-interactions and gravitational collapse. As such, they provide a distinction between quark (lepton) flavor-doublets in different families and also provide a way to distinguish between quarks (leptons) in the *same* flavor doublet.

To "completely" determine flavors[28] and families given the U-vectors of Fig. 4.1, we must either find the electric-charge vectors $\mathbf{Q}$ (i.e., $\mathbf{Q}_q$, $\mathbf{Q}_\ell$, $\mathbf{Q}_q^c$ and $\mathbf{Q}_\ell^c$) or, equivalently, their respective corresponding V-vectors, $\mathbf{V} = \mathbf{Q} - \mathbf{U}$. Having accomplished either of these tasks, we will have found all possible vector-triads ($\mathbf{Q}_q$ or $\mathbf{Q}_\ell$), $\mathbf{U}$, $\mathbf{V}$ or ($\mathbf{Q}_q^c$ or $\mathbf{Q}_\ell^c$), $\mathbf{U}$, $\mathbf{V}$. Thus, given $\mathbf{U}$, we need only solve eigenvalue-equations similar to Eqs. (2.15) and (2.16) for the electric-charge vectors and their components—the quark and lepton electric charges. In previous chapters, we treated these electric charges as being *known* from experiment (see Table I). In the following chapter they will be treated as *unknowns*.

4.2.4 Family Kinship—Preliminary Considerations

In the previous section, it was demonstrated that all fundamental fermions (quarks and leptons) in the *same* family have the *same* numerical $\mathbf{U}^2$-charges, i.e., $\mathbf{U}^2 = 0$ or ± 1. To ensure that 2-space charges are either conserved, or at most violated to first-order in most weak processes, it is necessary to *suppress* interfamily weak transitions. That is, "diagonal" weak transitions characterized by $\Delta \mathbf{U}^2 = 0$ and $\Delta B = \Delta L = 0$ are to be *preferred*, while "off-diagonal" weak transitions characterized by $\Delta \mathbf{U}^2 \neq 0$ and $\Delta B = \Delta L = 0$

4.2. THE SIX PHYSICALLY ACCEPTABLE U-VECTORS 49

are to be *suppressed*. We know from experiment (e.g., KM matrix elements) that this is so, but what is there about the 2-space geometry, if anything, that would "explain" this?

Let us begin by insisting, as we have from the outset, that the 2-space geometry is capable of providing a self-consistent description of *all* possible flavors of fundamental fermions.[29] Clearly, this can only happen if every flavor in Fig. 4.2 can be *distinguished* from every other flavor. And, this in turn means, among other things, that the quantum numbers S, S', t and t' must exhibit a strong tendency to be *separately* conserved. Hence, the suppression of interfamily weak-transitions is, to a certain extent, implicit in the 2-space description itself. Unfortunately, this stated tendency for suppression of interfamily transitions does not give us much insight. To gain a better geometric understanding of the situation, consider the 2-vector "selection-rules" that "govern" these processes.

Since individual flavors and flavor doublets are represented by vector-triads such as $(\mathbf{Q}, \mathbf{U}, \mathbf{V})$, where $\mathbf{Q} = \mathbf{U} + \mathbf{V}$, the vector-selection-rule $\Delta \mathbf{Q} = \Delta \mathbf{U} + \Delta \mathbf{V}$ applies to *all* conventional transitions. These include neutral-current processes as well as weak transitions that change electric charge. Furthermore, let us agree that all processes under consideration separately conserve B and L, which is equivalent to the vector-selection-rule $\Delta \mathbf{Q} = 0$. Then, for all such processes, we have the associated vector-selection-rule $\Delta \mathbf{U} = -\Delta \mathbf{V}$.

Now, in any transition, the associated vector-triad either remains the same or it changes. Suppose that the process under consideration is such that the vector-triad symmetry is unchanged, i.e., the vector-triad before the transition is *congruent* to the vector-triad after the transition. In this case, we would have the vector-selection-rule $\Delta \mathbf{U} = -\Delta \mathbf{V} = 0$, and interfamily transitions would be *excluded*. Hence, we conclude that *interfamily transitions must involve a change in the symmetry of the associated vector-triad.* Thus, the suppression of interfamily-transitions appears to be "equivalent" to the suppression of such symmetry changes (see also Sec. 5.4).

Family "kinship," which seems to be inherent to the 2-space description, will be described in more detail after other quantum numbers and charge labels for other fundamental-fermions have been introduced and fully described.

The extra families may not be as superfluous as they seem. Without them, there is no natural violation of time-reversal symmetry and, consequently, the universe would be balanced equally between matter and antimatter.

—S. L. Glashow

5. Effective "Quantization" of the Q_q and Q_ℓ Vector Components

Consider the case of quarks first (see Fig. 3.2). Using the new organizing principle stated in Sec. 1.2.6, the quark electric-charge vectors $\mathbf{Q}_q$ and $\mathbf{Q}_q^c$, treated as *unknowns*, are given, respectively, by Eqs. (1.12) and (1.13). To solve these eigenvalue equations it is necessary to determine $\mathbf{F}$ exactly. In other words, it is necessary to determine which sign (upper or lower) should be chosen in (1.21), and also the value to be assigned to the group parameter v.

To determine the sign, there is the requirement (see Fig. 3.2) that the "unknown" components of $\mathbf{Q}_q = \{q_1, q_2\}$, namely, q_1 and q_2 are global electric-charge labels having *opposite* signs when both q_1 and q_2 are nonzero. That is, the 2-vector $\mathbf{Q}_q = \{q_1, q_2\}$ may be said to "reside" in quadrant IV of the 2-space (see Figs. 3.1 and 3.2).

Given only (1.21) and the foregoing constraint, we find that the upper signs in (1.21) must be chosen. Hence, the "generalized fermion-number" $\mathbf{F}$ or $\mathbf{F}(v)$ can be represented by the real matrix

$$\mathbf{F} = \mathbf{F}(v) = \begin{pmatrix} \cosh v & \sinh v \\ -\sinh v & -\cosh v \end{pmatrix}, \qquad (5.1)$$

where the "quantized" values of the real dimensionless-parameter $v(-\infty \leq v \leq \infty)$ are yet to be determined.

5.1 New Labeling and Subscript Conventions

Using the parameter v to designate quarks and v' to designate leptons as suggested in Sec. 1.2.3, and dropping the matter-antimatter subscripts m and a on the scalar fermion-numbers, we can rewrite the basic eigenvalue equations applicable to matter ($f = f_m = +1$) and antimatter ($f = f_a = -1$)

5.1. NEW LABELING AND SUBSCRIPT CONVENTIONS

in a somewhat more useful and descriptive way as follows (primes designate lepton charges):

$$\mathbf{F}(v)\mathbf{Q}_q(v, f) = f\mathbf{Q}_q(v, f), \tag{5.2}$$

$$\mathbf{F}(v')\mathbf{Q}_\ell(v', f) = f\mathbf{Q}_\ell(v', f), \tag{5.3}$$

where

$$\mathbf{Q}_q(v, f) = \begin{pmatrix} q_1(v, f) \\ q_2(v, f) \end{pmatrix} \tag{5.4}$$

$$\mathbf{Q}_\ell(v', f) = \begin{pmatrix} q'_1(v', f) \\ q'_2(v', f) \end{pmatrix} \tag{5.5}$$

$$q_1(v, f) - q_2(v, f) = 1 \tag{5.6}$$

and

$$q'_1(v', f) - q'_2(v', f) = -1. \tag{5.7}$$

Notice that the subscripts 1 and 2 in Eqs. (5.4) through (5.7) consistently designate vector-component *positions* (hence coordinate-axes X_1 and X_2), respectively, for both matter and antimatter. That is, there is no reversal of subscripts 1 and 2 in going from matter to antimatter. The relation between this method of representing (and subscripting) electric-charge labels and that specified in Sec. 2.2 and Table I, is given by the following equations:

$$\left. \begin{array}{l} q_1(v, +1) = +q_1 \\ q_1(v, -1) = -q_2 \\ q_2(v, +1) = +q_2 \\ q_2(v, -1) = -q_1 \\ q'_1(v', +1) = +q'_1 \\ q'_1(v', -1) = -q'_2 \\ q'_2(v', +1) = +q'_2 \\ q'_2(v', -1) = -q'_1. \end{array} \right\} \tag{5.8}$$

The right-hand side of Eqs. (5.8) are the charges specified in Table I (treated here as *unknowns*), while the left-hand side of these equations are charges associated with the new, more general, subscripts. By solving the eigenvalue equations [i.e., Eqs. (5.2) and (5.3)] for the charges on the left-hand side of (5.8), the charges on the right-hand side of (5.8), which were formerly taken from experiment, will be determined.

Using (5.8) and the definitions of B and L discussed earlier [see the linear expressions in Eqs. (2.2) through (2.5)], we see that B can also be rewritten as

$$B = q_1(v, f) + q_2(v, f), \qquad (5.9)$$

while L can be rewritten as a similar *inverted* expression

$$B' = L = -q'_1(v', f) - q'_2(v', f). \qquad (5.10)$$

5.2 Effective "Quantization" of v and v'

Because the eigenvalues of $\mathbf{F}(v)$ are independent of the choice of the parameter v, Eqs. (5.1), (5.4) and (5.5) ensure that the eigenvectors of $\mathbf{F}(v)$ are functions of v (this will be demonstrated momentarily). Therefore, *the unknown dimensionless-parameter v determines the numerical values of the electric-charge labels q_1 and q_2*, which are the components of the vector $\mathbf{Q}_q$. This means that if we can somehow effectively "quantize" the parameter v, we will have "quantized" the electric-charge labels. It happens that only *two* choices for v are consistent with the standard-model description of the known first-family fractionally-charged quarks (characterized by v) and the known first-family integrally-charged leptons (characterized by v').

Given Eqs. (5.1) and (5.4), the eigenvector solutions to the eigenvalue equation (5.2) yield the following (quark) global electric-charge labels

$$q_1(v, f) = \frac{\sinh v}{e^v - f} \qquad (5.11)$$

and

$$q_2(v, f) = q_1(v, f) - 1. \qquad (5.12)$$

Similarly, the solutions of (5.3) yield the "inverted" (lepton) global electric-charge labels

$$q'_1(v', f) = \frac{-\sinh v'}{e^{v'} - f} \qquad (5.13)$$

5.2. EFFECTIVE "QUANTIZATION" OF v AND v'

and
$$q'_2(v', f) = q'_1(v', f) + 1. \tag{5.14}$$

These results can be verified by direct substitution of Eqs. (5.11) through (5.14) in the appropriate eigenvalue equation (5.2) or (5.3). Using these equations we can also easily determine, by direct substitution, that

$$e^v \mathbf{Q}_q^2 = f, \tag{5.15}$$

and by similar substitutions

$$e^{v'} \mathbf{Q}_\ell^2 = f, \tag{5.16}$$

where the global baryon-number $\mathbf{B} = \mathbf{Q}_q^2$ or $(\mathbf{Q}_q^c)^2$ or global lepton-number $B' = L = \mathbf{Q}_\ell^2$ or $(\mathbf{Q}_\ell^c)^2$ for an individual quark or lepton, respectively, is defined by Eqs. (5.9) and (5.10), respectively. Now let us determine how to choose the "quantized" parameters v and v' for quarks and leptons, respectively, by examining some of the properties of strong interactions.

5.2.1 Strong Interactions and the Parameters v and v'.

All fermions (antifermions), be they *composite* or *fundamental*, may be assigned the *same* scalar fermion-number $f = 1$ ($f = -1$).[16] This simple fact provides connections between the present formalism and the conventional strong-interactions that bind three quarks to form a *composite* fermion or, alternatively, "bind" one lepton to "form" a *fundamental* fermion.

From Eqs. (5.9), (5.10), (5.15) and (5.16), we know that for quarks

$$e^v B = f, \tag{5.17}$$

and for leptons

$$e^{v'} L = f. \tag{5.18}$$

Equations (5.17) and (5.18) demonstrate that e^v and $e^{v'}$ must play two different, but complementary, roles. First, it is clear that e^v ($e^{v'}$) *counts the number of quarks (leptons) carrying baryon (lepton) number B ($B' = L$) in a would-be strongly-bound composite-fermion carrying the overall scalar fermion-number f*. This means that e^v and $e^{v'}$ must be *positive* (i.e., both

v and $v' \geq 0$) *odd-integers* since the number of fundamental fermions carrying baryon number B or lepton number L, that make up any would-be composite fermion is, necessarily, an *odd* integer because of the spin-statistics theorem. Second, given Eqs. (5.9) and (5.10) for B and L, respectively, and noting that these equations are *linear* in the electric-charge labels (i.e., linear combinations of electric charges), *the positive odd integer e^v ($e^{v'}$), necessarily, counts the number of certain internal quark (lepton) states*, to be identified below with the strong-color multiplicity. For example, e^v counts the number of strong-color states (R, W, B) for quarks carrying the *same* electric charges. Thus, *the sum of the global electric-charges carried by the four flavors, in any quark-lepton family, has the appropriate value zero*, namely,

$$e^v B - e^{v'} L = 0. \tag{5.19}$$

This last result is essentially the condition of *global electric-charge quantization* [see (5.17), (5.18) and Ref. 4, p. 275]. Therefore, we find that the simplest possible choice of v' for electroweakly-interacting leptons, which cannot be strongly-bound into fermion (or boson) composites is, necessarily,

$$e^{v'} = 1, \tag{5.20}$$

for which $L = 1$ when $f = +1$. The simplest possible remaining choice of v for strongly-interacting quarks, which can be strongly bound into fermion composites is, necessarily,

$$e^v = 3, \tag{5.21}$$

for which $B = +1/3$ when $f = +1$. Equations (5.20) and (5.21) then determine the effectively-"quantized" values of v' and v, respectively. Therefore, they completely determine the matrices $\mathbf{F}(v')$ and $\mathbf{F}(v)$.

To paraphrase and extend these results, notice that the foregoing arguments evidently derive from certain "connections" between the global, internal 2-space and the local gauge-group $SU(3)_c$. That is, according to local $SU(3)_c$ and quantum chromodynamics (QCD), quarks are strong-color triplets,[4,5] i.e., they may be said to possess a strong-color multiplicity $M_c = 3$, while leptons are strong-color singlets,[4,5] i.e., they may be said to possess a strong-color multiplicity $M_c = 1$. This means that a positive (odd-integral) number $N = M_c$ of quarks, or leptons, will bind together strongly to form a composite fermion in a strong-color singlet state. That is, three quarks

5.2. EFFECTIVE "QUANTIZATION" OF v AND v'

($N = 3$) will bind together to form a strong-color singlet, while leptons have no strong interactions and thus occur only singly ($N = 1$).

Just as for individual fundamental fermions, the fermion number for any of these "composites" should be[16] $f = \pm 1$, where f is, necessarily, given by an expression of the general form

$$M_c \mathbf{Q}^2 = f. \tag{5.22}$$

Here, $M_c = N$, $\mathbf{Q}$ is $\mathbf{Q}_q$ or $\mathbf{Q}_\ell$ ($\mathbf{Q}_q^c$ or $\mathbf{Q}_\ell^c$) and $\mathbf{Q}^2 = B$ or L. Comparing (5.22) with Eqs. (5.15) and (5.16), we find that the global parameter v or v' is given by the "quantum condition":

$$v = \ln(M_c). \tag{5.23}$$

Hence, v and v' are known positive constants for all quarks ($M_c = 3$) and all leptons ($M_c = 1$), respectively. Note that these numbers are independent of particular local color-charges or flavor-indices since they depend *only* on the strong-color multiplicity M_c, which is a global color-independent parameter.

The substitution of (5.23) with $M_c = 3$ into Eqs. (5.11) and (5.12) correctly yields the experimentally observed numerical values for the first-family quark (antiquark) global electric-charge labels [see Table I and Eq. (5.8)], namely,

$$q_1(v, f) = \frac{(M_c^2 - 1)}{2M_c(M_c - f)}, \tag{5.24}$$

and $q_2(v, f)$, which is given by Eqs. (5.12) using (5.24).

The observed lepton (antilepton) global electric-charge labels are obtained from a similar "inverted" expression [see Table I and Eq. (5.8)], namely,

$$q'_1(v', f) = \frac{-(M_c^2 - 1)}{2M_c(M_c - f)}, \tag{5.25}$$

where the limit is taken as M_c approaches one and $q'_2(v', f)$ is given by (5.14) using (5.25) evaluated at the limit. The baryon [lepton] number for quarks [leptons] from Eq. (5.17) [Eq. (5.18)] is $B = f/3 = \pm 1/3$ [$B' = L = f = \pm 1$].

Equations (5.11) and (5.13) or, equivalently, Eqs. (5.24) and (5.25), respectively, are an unconventional way to describe or represent quark and lepton global electric-charge labels, even though they are numerically *identical* to the charges predicted by local $SU(5)$ and QCD. In $SU(5)$, the four

C-reversing colors, namely, the three strong-colors R, W, B and the weak-color G, each make "additive" contributions to the total electric-charge [see Refs. 4 and 5 and Eq. (2.1)]. But, in Eqs. (5.24) and (5.25) only the total *number* of strong-color charges, i.e., the global, color-independent parameter M_c and the fundamental unit of electric charge, which can be taken to be the weak-color charge $G = \pm 1$, enters. Similar arguments apply to the leptons.

5.3 An Interpretation of the Components of F

The general matrix **F** is C-reversing, which means that its components transform like "charges" [see Eq. (1.14)]. What are these charges?

Equations (5.8), (5.9), (5.10), (5.17) and (5.18), with $f = \pm 1$, demonstrate that

$$\left. \begin{array}{rcl} \cosh v & = & +[(q_1 + q_2) + (q_1 + q_2)^{-1}]/2 \\ \sinh v & = & -[(q_1 + q_2) - (q_1 + q_2)^{-1}]/2 \\ \cosh v' & = & -[(q_1' + q_2') + (q_1' + q_2')^{-1}]/2 \\ \sinh v' & = & +[(q_1' + q_2') - (q_1' + q_2')^{-1}]/2. \end{array} \right\} \quad (5.26)$$

Therefore, the components of the matrix **F** that describes quarks (leptons), are expressible as functions of the electric charges of the two quarks (leptons) in a flavor doublet. Clearly, the matrix $\mathbf{F}(v)$ [$\mathbf{F}(v')$], like these electric charges, is completely *independent* of the particular family in which the quark [lepton] happens to reside.

5.4 Symmetries and Asymmetries Associated With the Matrix F

Global flavor-doublet-preserving transformations (on vector-triads representing flavor doublets) exist and have the simplest possible nontrivial, abstract group structure. According to (1.10), the matrices $\mathbf{F}(v)$ and $\mathbf{F}(v')$ are generators of two-dimensional realizations of the finite two-element discrete group[33] Z_2, namely,

$$Z_2(v) = \{\mathbf{F}(v), \mathbf{I}_2\}, \qquad (5.27)$$

where $v = v$ or v'.

5.4. SYMMETRIES ASSOCIATED WITH THE MATRIX F

5.4.1 Symmetry Under F or Symmetry Class I

If the charge-vector triad representing a particular quark (antiquark) or lepton (antilepton) flavor-doublet *doesn't* change under one application of $\mathbf{F}(v)$ or $\mathbf{F}(v')$, respectively, it is obviously *symmetric* with respect to $\mathbf{F}(v)$ or $\mathbf{F}(v')$, respectively. These flavor doublets reside in what we call *Symmetry Class I* of $Z_2(v)$ or $Z_2(v')$, respectively. Hereinafter, this symmetry classification will be labeled $S(I)$. $S(I)$ will be shown to apply to first-family flavor-doublets or their corresponding antiparticles only.

5.4.2 Asymmetry Under F or Symmetry Class II

If the charge-vector triad representing a particular quark (antiquark) or lepton (antilepton) flavor-doublet *does* change under one application of $\mathbf{F}(v)$ or $\mathbf{F}(v')$, respectively, it is obviously *asymmetric* with respect to $\mathbf{F}(v)$ or $\mathbf{F}(v')$, respectively (see Ref. 34, p. 97). These flavor doublets reside in what we call *Symmetry Class II* of $Z_2(v)$ or $Z_2(v')$, respectively. Hereinafter, this symmetry classification will be labeled $S(II)$. $S(II)$ will be shown to apply to the second- and third-family flavor-doublets only.

Clearly, the symmetrical charge-vector triads in $S(I)$ are preserved under *both* $\mathbf{F}(v$ or $v')$ and the identity $\mathbf{F}^2(v$ or $v') = \mathbf{I}_2$, whereas the asymmetrical charge-vector triads in $S(II)$ are preserved *only* under the identity $\mathbf{F}^2(v$ or $v') = \mathbf{I}_2$. Hence, $S(I)$-triads transform under $\mathbf{F}(\mathbf{F}^2)$ "like" spin 0 states or ordinary 3D objects under 360° (720°) rotations in Euclidean 3-space. And, $S(II)$-triads transform under $\mathbf{F}(\mathbf{F}^2)$ "like" spin $\frac{1}{2}$ states or Möbius-strips under 360° (720°) rotations.

5.4.3 Symmetry "Evolution"

The symmetry-evolution principle (see Ref. 34, p. 161), which is closely associated with the second-law of thermodynamics, states:

> *For an isolated physical system the degree of symmetry cannot decrease as the system evolves, but either remains constant or increases.*

Suppose we naïvely treat the internal 2-space and its geometric "contents" as an *isolated* physical system (see Sec. 1.3.1). In particular, consider the symmetry "evolution" of *charge-vector triads* that represent both flavor-doublets of fundamental fermions and *individual* fundamental-fermions.

In Sec. 5.4.2, it was shown that vector-triads in $S(II)$ are *asymmetric* with respect to $\mathbf{F}$, while those in $S(I)$ are *symmetric* with respect to $\mathbf{F}$. Hence, $S(I)$ has a *higher* degree of symmetry than $S(II)$. This means that whenever an "isolated" fundamental fermion undergoes a weak *decay* its associated vector-triad and symmetry class [i.e., $S(I)$ or $S(II)$], will either remain the *same* (as in "diagonal" weak decays) or undergo a *change* (as in some "off-diagonal" weak decays). In particular, if we naïvely apply the symmetry-evolution principle to vector-triads associated with weak decays, these triads will generally "evolve" toward triads possessing the highest possible degree of symmetry, i.e., toward $S(I)$, namely,

$$S(II) \to S(II) \to S(I)$$

or

$$S(I) \to S(I).$$

Since this "evolution" is presumably always toward $S(I)$ for a triad, since this evolution is assumed to be associated with weak *decays* of individual fundamental-fermions, and since the fermions in such weak decays must proceed from an initial state of *higher* mass-energy to a final state of *lower* mass-energy, *the masses of $S(II)$-type fundamental fermions (involved in these decays) must be greater than the masses of $S(I)$-type fundamental fermions (involved in these same decays).*

Since there are only $S(I)$- or $S(II)$-type fundamental-fermions, the $S(I)$-type fundamental-fermions must correspond to the "ground" state. This implies that $S(I)$ describes flavor doublets in the *first* family, while $S(II)$ describes flavor doublets in the *second* and *third* families. Results presented in the next chapter will support these symmetry-based speculations.

Who ordered that?

—I. I. Rabi

6. A Hierarchy Among Families Revealed

If contact is to be made with experimental observations, each of the six physically acceptable **U**-vectors of Fig. 4.1 must be assigned to one of the six known quark, or lepton, flavor doublets (e.g., see Fig. 4.2). In this chapter, it will be demonstrated, among other things, that a unique flavor-doublet "raising"-matrix **R** exists and that this matrix automatically organizes the six physically acceptable **U**-vectors into a *hierarchy* analogous to the observed family hierarchy. In particular, the matrix **R** maps **U**-vectors associated with $-\sigma_x$ and $\pm\sigma_z$ into themselves in a hierarchical fashion.

The **U**-vectors of quadrant IV (see Footnote 32) that describe the three known quark flavor-doublets (see Figs. 3.1, 4.1 and 4.2) form the hierarchy: $\{1/2, -1/2\} \to \{1, 0\} \to \{0, -1\}$. Here, $\{1/2, -1/2\}$ describes the (u, d) flavor doublet, $\mathbf{R}\{1/2, -1/2\} = \{1, 0\}$ describes the (c, s) flavor doublet and $\mathbf{R}\{1, 0\} = \{0, -1\}$ describes the (t, b) flavor doublet. Similarly, the inverted **U**-vectors of quadrant II (see Footnote 32) that describe the three known lepton flavor-doublets (see Figs. 3.1, 4.1 and 4.2) form the hierarchy: $\{-1/2, 1/2\} \to \{-1, 0\} \to \{0, 1\}$. Here, $\{-1/2, 1/2\}$ describes the (e^-, ν_e) flavor doublet, $\mathbf{R}\{-1/2, 1/2\} = \{-1, 0\}$ describes the (μ^-, ν_μ) flavor doublet and $\mathbf{R}\{-1, 0\} = \{0, 1\}$ describes the (τ^-, ν_τ) flavor doublet.

6.1 Representation of Flavor Doublets by U-Vectors

Since there is supposed to be a one-to-one correspondence between **U**-vectors and flavor doublets, since the **U**-vectors describing quarks fall in quadrant IV of Fig. 4.1 (see Footnote 32 and Fig. 4.2), and since the (u, d), (c, s) and (t, b) quark flavor-doublets are known to fall in three *different* families (see Sec. 2.1), we propose that quark flavor-doublets be represented by the three **U**-vectors in quadrant IV of Fig. 4.1 (see Footnote 32 and Fig. 4.2) as

follows:
$$\begin{aligned}(u,d) &\Leftrightarrow \mathbf{U} = \{1/2, -1/2\} = \{T_3^+, T_3^-\} \\ (c,s) &\Leftrightarrow \mathbf{U} = \{1, 0\} = \{C, 0\} \\ (t,b) &\Leftrightarrow \mathbf{U} = \{0, -1\} = \{0, b\}.\end{aligned} \qquad (6.1)$$

The far right-hand side of these equations represent a possible interpretation of the **U**-vector components (see Fig. 4.2), and also provide some justification for these assignments. For example, the (u,d) assignment is justified by the conventional $SU(3)$-flavor assignments[26,27] for the third-component of global *isospin* of the first-family u and d quarks, namely, $T_3^+ = 1/2$ and $T_3^- = -1/2$, respectively. The remaining assignments in (6.1) imply that the *charm* $(C = +1)$ and *beauty* $(b = -1)$, carried by the c and b quarks, respectively, is given by the nonzero components of the corresponding **U**-vectors (see Fig. 4.2).[35] In general, different *names* are assigned to different **U**-vector components because they are taken to be *separately* conserved in *strong-* and *electromagnetic-interactions* and some neutral-current *weak-interactions*.

Because of the effect of the matter-antimatter reflection $-\boldsymbol{\sigma}_x$ on the **U**-vectors of (6.1), namely, [see Eq. (1.17)],
$$\begin{aligned}-\boldsymbol{\sigma}_x\{1/2, -1/2\} &= \{1/2, -1/2\} \\ -\boldsymbol{\sigma}_x\{1, 0\} &= \{0, -1\} \\ -\boldsymbol{\sigma}_x\{0, -1\} &= \{1, 0\},\end{aligned} \qquad (6.2)$$

these same three **U**-vectors, in a different *order*, must represent the three corresponding *antiquark* flavor-doublets. In particular, consistency between Eqs. (6.1) and (6.2) forces us to extend the identifications made in (6.1) as follows:
$$\begin{aligned}(u,d),(\bar{d},\bar{u}) &\Leftrightarrow \mathbf{U} = \{1/2, -1/2\}, & \mathbf{U}^2 &= 0, & S(I) \\ (c,s),(\bar{b},\bar{t}) &\Leftrightarrow \mathbf{U} = \{1, 0\}, & \mathbf{U}^2 &= +1, & S(II_1) \\ (t,b),(\bar{s},\bar{c}) &\Leftrightarrow \mathbf{U} = \{0, -1\}, & \mathbf{U}^2 &= -1, & S(II_2).\end{aligned} \qquad (6.3)$$

Thus, the *same* **U**-vector component can represent *different* quantum numbers (see Fig. 4.2). In (6.3) the symbols $S(I)$ and $S(II_1)$ or $S(II_2)$ designate different types of *symmetry* (under the matrix **F**) of vector-triads that represent flavor doublets (see Secs. 5.4.1 through 5.4.3 and Appendix C). Note that the subscript 1 or 2 [associated with $S(II_1)$ or $S(II_2)$] refers to the fact that the corresponding **U**-vector lies in the one-dimensional invariant vector-spaces associated with $+\boldsymbol{\sigma}_z$ or $-\boldsymbol{\sigma}_z$, namely, the X_1 or X_2 coordinate axes, respectively.

6.2. THE FAMILY HIERARCHY AND THE MATRIX R

Similarly, lepton and antilepton flavor-doublets are represented by the corresponding *inverted* **U**-vectors of quadrant II as follows:

$$
\begin{array}{llll}
(e^-, \nu_e), (\bar{\nu}_e, e^+) & \Leftrightarrow \mathbf{U} = \{-1/2, 1/2\}, & \mathbf{U}^2 = 0, & S(I) \\
(\mu^-, \nu_\mu), (\bar{\nu}_\tau, \tau^+) & \Leftrightarrow \mathbf{U} = \{-1, 0\}, & \mathbf{U}^2 = +1, & S(II_1) \\
(\tau^-, \nu_\tau), (\bar{\nu}_\mu, \mu^+) & \Leftrightarrow \mathbf{U} = \{0, 1\}, & \mathbf{U}^2 = -1, & S(II_2).
\end{array}
\right\} \quad (6.4)
$$

In particular, **U**-vectors describing flavor-doublets of quarks (antiquarks) and leptons (antileptons) that happen to be in the *same* family [i.e., (u, d, e^-, ν_e), (c, s, μ^-, ν_μ) or (t, b, τ^-, ν_τ)], are the *additive inverses* of one another (see Figs. 3.1, 4.1 and 4.2).

6.2 The Family Hierarchy and the Matrix R

Given the six physically-acceptable **U**-vectors of Fig. 4.1, together with their assignments to flavor- and antiflavor-doublets [see Eqs. (6.3) and (6.4) and Fig. 4.2], it is easily demonstrated that a "transformation" represented by the 2×2 matrix

$$\mathbf{R} = \begin{pmatrix} 0 & -2 \\ -1 & -1 \end{pmatrix}, \quad (6.5)$$

"acting" on **U**-vectors *alone*, serves as a global flavor-doublet "raising"-matrix for all quark and lepton flavor-doublets, while the 2×2 matrix (see Appendix B)

$$\mathbf{R}^c = (-\boldsymbol{\sigma}_x)\mathbf{R}(-\boldsymbol{\sigma}_x) \quad (6.6)$$

or

$$\mathbf{R}^c = \begin{pmatrix} -1 & -1 \\ -2 & 0 \end{pmatrix}, \quad (6.7)$$

serves as a global flavor-doublet "raising"-matrix for all antiquark and antilepton flavor-doublets. That is,

$$
\left.
\begin{array}{l}
\mathbf{R}\{\pm 1/2, \mp 1/2\} = \{\pm 1, 0\} \\
\mathbf{R}\{\pm 1, 0\} = \{0, \mp 1\} \\
\mathbf{R}^c\{\pm 1/2, \mp 1/2\} = \{0, \mp 1\} \\
\mathbf{R}^c\{0, \mp 1\} = \{\pm 1, 0\}
\end{array}
\right\} \quad (6.8)
$$
.

Clearly, the multiplicative-inverses of these 2 × 2 matrices, namely,

$$\mathbf{R}^{-1} = \begin{pmatrix} \frac{1}{2} & -1 \\ -\frac{1}{2} & 0 \end{pmatrix} \tag{6.9}$$

and

$$(\mathbf{R}^c)^{-1} = \begin{pmatrix} 0 & -\frac{1}{2} \\ -1 & \frac{1}{2} \end{pmatrix}, \tag{6.10}$$

are the corresponding flavor-doublet "lowering"-matrices. Let us summarize the foregoing observations.

The effect of the flavor-doublet raising-matrices $\mathbf{R}$ and $\mathbf{R}^c$ on the various flavor doublets (i.e., the associated U-vectors representing these flavor doublets), and their associated symmetry classes $S(I)$, $S(II_1)$ and $S(II_2)$, can be illustrated in symbolic form in the following way:

$$\left. \begin{array}{c} S(I) \overset{\mathbf{R}}{\Rightarrow} S(II_1) \overset{\mathbf{R}}{\Rightarrow} S(II_2) \\ (u,d) \Rightarrow (c,s) \Rightarrow (t,b) \\ (e^-, \nu_e) \Rightarrow (\mu^-, \nu_\mu) \Rightarrow (\tau^-, \nu_\tau) \\ \underbrace{\qquad\qquad \mathbf{R}^2 \qquad\qquad}_{} \end{array} \right\} \tag{6.11}$$

and

$$\left. \begin{array}{c} S(I) \overset{\mathbf{R}^c}{\Rightarrow} S(II_2) \overset{\mathbf{R}^c}{\Rightarrow} S(II_1) \\ (\bar{d}, \bar{u}) \Rightarrow (\bar{s}, \bar{c}) \Rightarrow (\bar{b}, \bar{t}) \\ (\bar{\nu}_e, e^+) \Rightarrow (\bar{\nu}_\mu, \mu^+) \Rightarrow (\bar{\nu}_\tau, \tau^+). \\ \underbrace{\qquad\qquad (\mathbf{R}^c)^2 \qquad\qquad}_{} \end{array} \right\} \tag{6.12}$$

It is significant that the third-family flavor-doublets in (6.11), namely, (t,b) and (τ^-, ν_τ), as represented by U-vectors, cannot be "raised" by $\mathbf{R}$ nor can the first-family flavor-doublets (u,d) and (e^-, ν_e) be "lowered" by $\mathbf{R}^{-1}$, since these operations invariably lead to U-vectors that are *not* among those allowed [see Eqs. (4.12) and (4.13) and Fig. 4.1].

While the foregoing properties of matrices like $\mathbf{R}$ are just another way of indicating that there are only *three* possible quark or lepton flavor-doublets, they have other implications including a basis for family membership of flavor doublets and a family hierarchy. Relations (6.11) and (6.12) illustrate that

6.3. AN INTERPRETATION OF THE COMPONENTS OF R

a particular quark or lepton flavor-doublet is to be placed in the same quark-lepton family (i.e., they are "kin") if, and only if, the **U**-vectors describing these flavor doublets transform in the same way under **R** and $\mathbf{R}^{-1}$.

This requirement automatically reveals a *hierarchy* among families that is analogous to the observed mass-hierarchy of families (see Sec. 2.1). In particular, the (t, b) and (τ^-, ν_τ) flavor-doublets, neither of which can be raised by **R**, necessarily, fall in the third quark-lepton family at the "top" of the hierarchy, while the (u, d) and (e^-, ν_e) flavor-doublets, neither of which can be lowered by $\mathbf{R}^{-1}$, necessarily, fall in the first quark-lepton family at the "bottom" of the hierarchy. Finally, the (c, s) and (μ^-, ν_μ) flavor-doublets, both of which can be raised by **R** or lowered by $\mathbf{R}^{-1}$, necessarily, fall in the second quark-lepton family lying in an intermediate position, i.e., they are "half-way" up the hierarchy.

The foregoing relationships, which are suggested by the 2-space algebra and geometry, strongly imply that the *names, symbols*, and *global charge-like quantum numbers* that we have assigned to the various fundamental-fermion flavors correspond closely to their conventional standard-model counterparts, especially, in the case of quarks (e.g., see Fig. 4.2).

6.3 An Interpretation of the Components of R

The matrix **R** is required to *raise* the first-family vector $\mathbf{U} = \{1/2, -1/2\}$ to a vector of the form [see Eq. (4.1)]

$$\mathbf{R}\{1/2, -1/2\} = \{a, a - 1\}, \tag{6.13}$$

where $a \neq \frac{1}{2}$. If **R** were C-invariant then

$$\mathbf{R} = \mathbf{R}^c, \tag{6.14}$$

and **R** would have the general form (see Appendix B)

$$\mathbf{R} = \begin{pmatrix} c & d \\ d & c \end{pmatrix}. \tag{6.15}$$

Then

$$\mathbf{R}\{1/2, -1/2\} = 1/2(c - d)\{1, -1\}, \tag{6.16}$$

where $c - d \neq 1$ since $a \neq \frac{1}{2}$. Clearly, (6.16) is not of the required form (6.13) for a general U-type vector. Hence, **R** is *not* C-invariant. If **R** is not C-invariant it could be C-reversing. In this case, each of the four scalar-components of **R** must transform like C-reversing charges.

Assume that **R** is C-reversing, and that the charge-like components of **R** are functions of four physically-observable charges carried by the four flavors in a fundamental-fermion family. Since **R** is supposed to operate on the first two families in the same way [see Eqs. (6.11) and (6.12)], these charge-like components will, necessarily, be independent of which family is involved.

The only known charges that fit this general description are the four *electric* charges corresponding to the four fundamental-fermions in any given family. Thus, the components of **R**, like those of **F**, may depend on electric-charge labels (see Sec. 5.3).

To find such a matrix **R**, compare the transformation of the four components of **R** under $-\sigma_x$ with the transformation of the four electric-charges in any family under $-\sigma_x$. The transformation of **R** under $-\sigma_x$ [see Eq. (6.6)] is such that the diagonal (off-diagonal) elements of **R** are *exchanged*. But, we also know that the four electric-charges in any family transform under $-\sigma_x$, as described by Eqs. (2.8) through (2.11), namely,

$$\left. \begin{array}{rcl} q_1 & \leftrightarrow & -q_2 \\ q_2 & \leftrightarrow & -q_1 \\ q'_1 & \leftrightarrow & -q'_2 \\ q'_2 & \leftrightarrow & -q'_1. \end{array} \right\} \tag{6.17}$$

If the four components of **R** were simply proportional to the four quark-lepton electric charges in a family, consistency between Eqs. (6.17) and (6.6) would require quark (lepton) electric-charges to appear as diagonal (off-diagonal) elements of **R**, or vice versa.

There are many matrices that fit this general description, but only one that is consistent with the requirement (6.13). That is, if the components of **R** are proportional to electric charges, **R** must be of the form

$$\mathbf{R} = \begin{pmatrix} -q'_2 & -3q_1 \\ +3q_2 & +q'_1 \end{pmatrix}, \tag{6.18}$$

where it is necessary to assume color-weighted quark and lepton electric-charges (i.e., the strong-color multiplicity or "weight" $M_c = 3$ for quarks and $M_c = 1$ for leptons). When this is done,

$$\mathbf{R}\{1/2, -1/2\} = 1/2\{(3q_1 - q'_2), (3q_2 - q'_1)\}, \tag{6.19}$$

6.3. AN INTERPRETATION OF THE COMPONENTS OF **R**

which has the correct form (6.13), provided

$$\frac{3}{2}(q_1 - q_2) + \frac{1}{2}(q'_1 - q'_2) = 1. \tag{6.20}$$

But, we know from Eqs. (3.1) and (3.2) that the fundamental unit of electric charge is

$$(q_1 - q_2) = -(q'_1 - q'_2) = 1, \tag{6.21}$$

which means that (6.20) is satisfied and, therefore, that **R** in (6.18) is consistent with (6.13), as required. Finally, substituting the numerical quark and lepton electric-charges from Table I in (6.18) the result is (6.5).

6.4 The Four Main Kinds of 2-Space Transformations

In Table II we summarize the four main kinds of global *symbolic* transformations characterizing the internal 2-space description of flavor doublets and families.

Table II. Summary of Global Transformations of Flavor Doublets. All transformations correspond to operations defined on the two-dimensional non-Euclidean linear vector-space described in Table I and depicted in Figs. 3.1, 3.2, 4.1 and 4.2. All transformations act directly on charge-vectors representing flavors or flavor-doublets rather than on flavor eigenstates directly. In particular, $\mathbf{F} = \mathbf{F}(v)$ or $\mathbf{F}(v')$, $-\boldsymbol{\sigma}_x$ and the inversion $\mathbf{I} = -\mathbf{I}_2$ act on each vector of any charge-vector triad ($\mathbf{Q}_q$ or $\mathbf{Q}_\ell$, $\mathbf{U}$, $\mathbf{V}$) representing a flavor doublet, whereas $\mathbf{R}$ acts only on the U-vector associated with a particular flavor-doublet. $\mathbf{F}(v)$ or $\mathbf{F}(v')$ is a matrix-generalization of the fermion number that also serves as a flavor-doublet-preserving symmetry transformation. $\mathbf{R}$ is a flavor-doublet-raising transformation and the matrix $-\boldsymbol{\sigma}_x$ plays a role analogous to charge-conjugation $\mathbf{C}$. Finally, the inversion $\mathbf{I}$ and a parameter v, taken together, serve to transform quark ($v = v$) flavor-doublets into lepton ($v = v'$) flavor-doublets and vice versa.

The matter-antimatter transformation $-\boldsymbol{\sigma}_x$; $(-\boldsymbol{\sigma}_x)^2 = \mathbf{I}_2$:

$-\boldsymbol{\sigma}_x(u,d) = (\bar{d}, \bar{u})$ $\quad -\boldsymbol{\sigma}_x(c,s) = (\bar{s}, \bar{c})$ $\quad -\boldsymbol{\sigma}_x(t,b) = (\bar{b}, \bar{t})$
$-\boldsymbol{\sigma}_x(e^-, \nu_e) = (\bar{\nu}_e, e^+)$ $\quad -\boldsymbol{\sigma}_x(\mu^-, \nu_\mu) = (\bar{\nu}_\mu, \mu^+)$ $\quad -\boldsymbol{\sigma}_x(\tau^-, \nu_\tau) = (\bar{\nu}_\tau, \tau^+)$

Flavor-doublet-raising transformations $\mathbf{R}$ and $\mathbf{R}^2$:

$\mathbf{R}(u,d) = (c,s)$ $\quad \mathbf{R}(c,s) = (t,b)$ $\quad \mathbf{R}^2(u,d) = (t,b)$
$\mathbf{R}(e^-, \nu_e) = (\mu^-, \nu_\mu)$ $\quad \mathbf{R}(\mu^-, \nu_\mu) = (\tau^-, \nu_\tau)$ $\quad \mathbf{R}^2(e^-, \nu_e) = (\tau^-, \nu_\tau)$

Flavor-doublet-preserving symmetry transformations $\mathbf{F}$ and ($\mathbf{F}^2 = \mathbf{I}_2$):

$\quad\quad\quad S(I) \quad\quad\quad\quad\quad\quad S(II_1) \quad\quad\quad\quad\quad\quad S(II_2)$
$\mathbf{F}(u,d) = (u,d) \quad\quad \mathbf{F}(c,s) \neq (c,s) \quad\quad \mathbf{F}(t,b) \neq (t,b)$
$\mathbf{F}^2(u,d) = (u,d) \quad\quad \mathbf{F}^2(c,s) = (c,s) \quad\quad \mathbf{F}^2(t,b) = (t,b)$
$\mathbf{F}(e^-, \nu_e) = (e^-, \nu_e) \quad \mathbf{F}(\mu^-, \nu_\mu) \neq (\mu^-, \nu_\mu) \quad \mathbf{F}(\tau^-, \nu_\tau) \neq (\tau^-, \nu_\tau)$
$\mathbf{F}^2(e^-, \nu_e) = (e^-, \nu_e) \quad \mathbf{F}^2(\mu^-, \nu_\mu) = (\mu^-, \nu_\mu) \quad \mathbf{F}^2(\tau^-, \nu_\tau) = (\tau^-, \nu_\tau)$

Quark$\Leftrightarrow$Lepton transformations (Inversion $\mathbf{I} = -\mathbf{I}_2$ and $v \to v' = 0$ or $v' \to v$):

$\mathbf{I}(u,d) = (e^-, \nu_e) \quad\quad\quad \mathbf{I}(e^-, \nu_e) = (u,d)$
$\quad\quad v \to v' \quad\quad\quad\quad\quad\quad\quad v' \to v$
$\mathbf{I}(c,s) = (\mu^-, \nu_\mu) \quad\quad\quad \mathbf{I}(\mu^-, \nu_\mu) = (c,s)$
$\quad\quad v \to v' \quad\quad\quad\quad\quad\quad\quad v' \to v$
$\mathbf{I}(t,b) = (\tau^-, \nu_\tau) \quad\quad\quad \mathbf{I}(\tau^-, \nu_\tau) = (t,b)$
$\quad\quad v \to v' \quad\quad\quad\quad\quad\quad\quad v' \to v$

Of all principles of physics, the laws of conservation of charge, lepton number, baryon number, mass and angular momentum are among the most firmly established. Yet with gravitational collapse the content of these conservation laws also collapses.

—C. W. Misner, K. S. Thorne & J. A. Wheeler

7. Flavor Eigenstates and Their Mutually-Commuting Charge Labels

Using the six **U**-vectors illustrated in Fig. 4.1, and the identifications made in Fig. 4.2 and Eqs. (6.3) and (6.4), we can now tabulate all of the charge-vectors, scalar-components and scalar-products, associated with fundamental fermion flavors, flavor-doublets and families (see Tables III through VII). These tabulated results constitute the essence of our *ex post facto* predictions regarding families. In Table III quark and antiquark properties are listed, and in Table IV lepton and antilepton properties are listed.[35,36]

Note that in Table IV, μ^- and ν_μ and their antiparticles each carry *zero* global generalized-hypercharges. That is, the two components of the associated **V**-vector are *both* zero. Hence, **V** is "null" and has zero length-square (i.e., $\mathbf{V}^2 = 0$). But, this is *not* a trivial situation, as it would be in a Euclidean space, because in this non-Euclidean space such a "null" **V**-vector is easily distinguished from other **V**-vectors having *nonzero* components (nonzero global generalized-hypercharges) and $\mathbf{V}^2 = 0$.

The conserved, global flavor-defining charge-labels or quantum numbers carried by fundamental-fermion flavor, or mass, eigenstates, include "up"-"down"-type *flavor-symmetric*, and "up"-"down"-type *flavor-asymmetrical* charge labels. The "up"-"down"-type flavor-asymmetrical charge-labels, which distinguish between "up"- and "down"-type flavors (i.e., they are different for "up"- and "down"-type flavors within a given flavor-doublet) include: quark (lepton) *electric charge* labels q_1 and q_2 (q'_1 and q'_2), the *third-component of global isospin* T_3 (T'_3), *charm* C (C') and *beauty* b (b').[35,36]

The "up"-"down"-type flavor-symmetric charge-labels, which do not distinguish between "up"-"down"-type flavors (i.e., they are the same for "up"- and "down"- type flavors within a given flavor-doublet) include: *baryon (lepton) number* B ($B' = L$), *strangeness* S (S'), and *truth* t (t').[35,36] Using the foregoing charge labels, we also have the "up"-"down"-type flavor-symmetric global "*generalized-hypercharge*" labels $Y = B + S + t$ ($Y' = B' + S' + t'$).

Table III. Quadrant IV Global Geometric Quantities and Quarks.

Charge vectors, in column-vector form, are $\mathbf{U} = \{a, a-1\}$, $\mathbf{V} = \{Y/2, Y/2\}$ and $\mathbf{Q}_q(v, f) = \mathbf{U} + \mathbf{V}$ or $\mathbf{Q}_q(v, f) = \{q_1(v, f), q_2(v, f)\}$. Charge scalars are $\mathbf{Q}_q^2(v, f) = B = q_1^2(v, f) - q_2^2(v, f)$ or $\mathbf{Q}_q^2(v, f) = fe^{-v}$; $\mathbf{U} \bullet \mathbf{V} = Y/2$ and $\mathbf{Q}_q^2(v, f) = \mathbf{U}^2 + 2\mathbf{U} \bullet \mathbf{V} + \mathbf{V}^2$ or $S + t = Y - B$; $S + t = -\mathbf{U}^2 = -2a + 1$. Here, $f = +1$ describes quark flavor-doublets and $f = -1$ describes antiquark flavor-doublets. The parameter $v = \ln M_c$, where M_c is the strong-color multiplicity of quarks, namely, $M_c = 3$. The symbols $S(I)$ and $S(II_1)$ or $S(II_2)$ designate different types of *symmetry*, of vector-triads $(\mathbf{Q}_q, \mathbf{U}, \mathbf{V})$ representing flavor-doublets (see Secs. 5.4.1 through 5.4.3 and Appendix C), under the matrix transformation $\mathbf{F}$. Note that the subscript 1 or 2, associated with $S(II_1)$ or $S(II_2)$, respectively, refers to the fact that the associated $\mathbf{U}$-vector lies in the one-dimensional invariant vector-spaces associated with the X_1 or X_2 coordinate-axes, respectively.

		U	V	$Q_q(v,f)$	B	$S+t$	
u	$S(I)$	$T_3 = +1/2$	$+1/6$	$+2/3$	$+1/3$	$(S=0, t=0)$,	0
d		$T_3 = -1/2$	$+1/6$	$-1/3$			
c	$S(II_1)$	$C = +1$	$-1/3$	$+2/3$	$+1/3$	$(S=-1, t=0)$,	-1
s		0	$-1/3$	$-1/3$			
t	$S(II_2)$	0	$+2/3$	$+2/3$	$+1/3$	$(S=0, t=+1)$,	$+1$
b		$b = -1$	$+2/3$	$-1/3$			
$\bar{d}$	$S(I)$	$T_3 = +1/2$	$-1/6$	$+1/3$	$-1/3$	$(S=0, t=0)$,	0
$\bar{u}$		$T_3 = -1/2$	$-1/6$	$-2/3$			
$\bar{s}$	$S(II_2)$	0	$+1/3$	$+1/3$	$-1/3$	$(S=+1, t=0)$,	$+1$
$\bar{c}$		$C = -1$	$+1/3$	$-2/3$			
$\bar{b}$	$S(II_1)$	$b = +1$	$-2/3$	$+1/3$	$-1/3$	$(S=0, t=-1)$,	-1
$\bar{t}$		0	$-2/3$	$-2/3$			

Table IV. Quadrant II Global Geometric Quantities and Leptons.

Charge vectors, in column-vector form, are $\mathbf{U} = \{-a, -a+1\}$, $\mathbf{V} = \{-Y'/2, -Y'/2\}$ and $\mathbf{Q}_\ell(v', f) = \mathbf{U} + \mathbf{V}$ or $\mathbf{Q}_\ell(v', f) = \{q'_1(v', f), q'_2(v', f)\}$. Charge scalars are $\mathbf{Q}^2_\ell(v', f) = B' = L = q'^2_1(v', f) - q'^2_2(v', f)$ or $\mathbf{Q}^2_\ell(v', f) = fe^{-v'}$; $\mathbf{U} \bullet \mathbf{V} = Y'/2$ and $\mathbf{Q}^2_\ell(v', f) = \mathbf{U}^2 + 2\mathbf{U} \bullet \mathbf{V} + \mathbf{V}^2$ or $S' + t' = Y' - B'$; $S' + t' = -\mathbf{U}^2 = -2a + 1$. Here, $f = +1$ describes lepton flavor-doublets and $f = -1$ describes antilepton flavor-doublets. The parameter $v' = \ln M_c$, where M_c is the strong-color multiplicity of leptons, namely, $M_c = 1$. The symbols $S(I)$ and $S(II_1)$ or $S(II_2)$ designate different types of *symmetry*, of vector-triads $(\mathbf{Q}_\ell, \mathbf{U}, \mathbf{V})$ representing flavor-doublets (see Secs. 5.4.1 through 5.4.3 and Appendix C), under the matrix transformation $\mathbf{F}$. Note that the subscript 1 or 2, associated with $S(II_1)$ or $S(II_2)$, respectively, refers to the fact that the associated $\mathbf{U}$-vector lies in the one-dimensional invariant vector-spaces associated with the X_1 or X_2 coordinate-axes, respectively.

		U	V	$\mathbf{Q}_\ell(v', f)$	B'	$S' + t'$	
e^-	$S(I)$	$T'_3 = -1/2$	$-1/2$	-1	$+1$	$(S' = 0, t' = 0),$	0
ν_e		$T'_3 = +1/2$	$-1/2$	0			
μ^-	$S(II_1)$	$C' = -1$	0	-1	$+1$	$(S' = -1, t' = 0),$	-1
ν_μ		0	0	0			
τ^-	$S(II_2)$	0	-1	-1	$+1$	$(S' = 0, t' = +1),$	$+1$
ν_τ		$b' = +1$	-1	0			
$\bar{\nu}_e$	$S(I)$	$T'_3 = -1/2$	$+1/2$	0	-1	$(S' = 0, t' = 0),$	0
e^+		$T'_3 = +1/2$	$+1/2$	$+1$			
$\bar{\nu}_\mu$	$S(II_2)$	0	0	0	-1	$(S' = +1, t' = 0),$	$+1$
μ^+		$C' = +1$	0	$+1$			
$\bar{\nu}_\tau$	$S(II_1)$	$b' = -1$	$+1$	0	-1	$(S' = 0, t' = -1),$	-1
τ^+		0	$+1$	$+1$			

7. FLAVOR EIGENSTATES

Table V. The Mutually-Commuting Flavor-Defining Global Charge Labels or Quantum Numbers and Flavor, or Mass, Eigenstates (ket Vectors) for Quarks.

		$\mid q_1$ or q_2,	T_3,	C,	b,	B,	S,	t $\rangle$
$S(I)$	$u = \mid$	$+2/3$,	$+1/2$,	0,	0,	$+1/3$,	0,	0 $\rangle$
	$d = \mid$	$-1/3$,	$-1/2$,	0,	0,	$+1/3$,	0,	0 $\rangle$
$S(II_1)$	$c = \mid$	$+2/3$,	0,	$+1$,	0,	$+1/3$,	-1,	0 $\rangle$
	$s = \mid$	$-1/3$,	0,	0,	0,	$+1/3$,	-1,	0 $\rangle$
$S(II_2)$	$t = \mid$	$+2/3$,	0,	0,	0,	$+1/3$,	0,	$+1$ $\rangle$
	$b = \mid$	$-1/3$,	0,	0,	-1,	$+1/3$,	0,	$+1$ $\rangle$

Table VI. The Mutually-Commuting Flavor-Defining Global Charge Labels or Quantum Numbers and Flavor, or Mass, Eigenstates (ket Vectors) for Leptons.

		$\mid q_1'$ or q_2',	T_3',	C',	b',	$B' = L$,	S',	t' $\rangle$
$S(I)$	$e^- = \mid$	-1,	$-1/2$,	0,	0,	$+1$,	0,	0 $\rangle$
	$\nu_e = \mid$	0,	$+1/2$,	0,	0,	$+1$,	0,	0 $\rangle$
$S(II_1)$	$\mu^- = \mid$	-1,	0,	-1,	0,	$+1$,	-1,	0 $\rangle$
	$\nu_\mu - \mid$	0,	0,	0,	0,	$+1$,	-1,	0 $\rangle$
$S(II_2)$	$\tau^- = \mid$	-1,	0,	0,	0,	$+1$,	0,	$+1$ $\rangle$
	$\nu_\tau = \mid$	0,	0,	0,	$+1$,	$+1$,	0,	$+1$ $\rangle$

In summary, the (global) flavor, or mass, eigenstates (simultaneous flavor-eigenstates) describing quarks are represented by the ket-vectors

$$|q_1 \text{ or } q_2, T_3, C, b, B, S, t\rangle, \tag{7.1}$$

where each of the charge labels defining the kets are taken to be conserved, mutually-commuting (compatible) observables.

7.1. THE NEW GLOBAL CHARGE RELATIONS

The (global) flavor, or mass, eigenstates (simultaneous flavor-eigenstates) describing leptons are represented by the ket-vectors

$$|q'_1 \text{ or } q'_2, T'_3, C', b', B' = L, S', t'\rangle, \qquad (7.2)$$

where each of the charge labels defining the kets are taken to be conserved, mutually-commuting (compatible) observables.

The nonzero values of the mutually-commuting (simultaneously observable) flavor-defining charge labels or quantum numbers corresponding to a particular flavor, or mass, eigenstate can be read off from Tables III and IV. But, for the reader's convenience we have listed these charge labels, and also provided the associated particle-states in Tables V and VI. The corresponding antiparticle states can be obtained from any of these states by simply changing the signs of all global flavor-defining charge labels. Table VII lists the numerical values of most of the 2-scalar "quantum numbers" or charges associated with quarks and leptons.

7.1 The new Global Charge Relations

The generalized Gell-Mann–Nishijima-like global charge-relation that has been found for quarks [see Table III and Eqs. (4.7) and (4.8)] is [Q_q(global) = q_1 or q_2]

$$Q_q(\text{global}) = T_3 + \frac{Y}{2} + C + b. \qquad (7.3)$$

The global generalized-hypercharge label Y in (7.3) is given by (4.16).

That Eqs. (7.3) and (4.16) are reasonable global charge-relations for quarks can be demonstrated by first setting the charge labels t and b in these equations equal to zero to eliminate the third quark flavor-doublet. This results in the charge relation advocated by Huang for the first two quark flavor-doublets, namely, the u, d, c and s quarks.[36] Huang's charge relation was designed to maintain the selection-rules governing a change of strangeness—as required in the Glashow-Iliopoulos-Maiani (GIM) mechanism.[37] By setting the charge labels t, b and C equal to zero in Eqs. (7.3) and (4.16), we recover the familiar Gell-Mann–Nishijima charge-relation (1.30) and the global hypercharge (1.31), respectively, for the u, d and s quarks.[26,27]

Table VII. Quantized C-Reversing 2-Scalars. Most of the effectively "quantized" C-reversing 2-scalar "charges" that have been found to be associated with ("carried by") quarks and leptons are listed below. The definitions of these quantities are to be found in the text, in other tables, and in the appendices. Note that the symbol Q (or Q') stands for the electric charges of the fundamental fermions (quark charges are $\pm 2/3, \pm 1/3$ and lepton charges are $0, \pm 1$), while **Q** is a charge-vector (see text for clarification). The primes indicate lepton charge-labels. Note that all of these 2-scalar charges are also assumed to be Lorentz 4-scalars.

Quantized "charges"	Numerical values
$\mathbf{Q}_q^2, \mathbf{Q}_\ell^2$	$\pm 1/3, \pm 1$
$\mathbf{U}^2$	$0, \pm 1$
S or S'	$0, \pm 1$
t or t'	$0, \pm 1$
$\mathbf{V}^2$	0
$\mathbf{U} \bullet \mathbf{V}$	$0, \pm 1/6, \pm 1/3, \pm 1/2, \pm 2/3, \pm 1$
Q or Q'	$0, \pm 1/3, \pm 2/3, \pm 1$
C or C'	$0, \pm 1$
b or b'	$0, \pm 1$
T_3 or T_3'	$0, \pm 1/2$

The "inverted" generalized Gell-Mann–Nishijima-like global charge-relation that has been found for leptons [see Table IV and Eqs. (4.9) and (4.10)] is $[Q_\ell(\text{global}) = q_1'$ or $q_2']$

$$Q_\ell(\text{global}) = T_3' - \frac{Y'}{2} + C' + b'. \tag{7.4}$$

The global generalized-hypercharge label Y' in (7.4) is given by (4.17).

7.2 Global Conservation Laws

The charge relations (7.3) and (7.4) imply that, under certain conditions, similar relations will hold for conserved currents associated with these global,

7.2. GLOBAL CONSERVATION LAWS

conserved charge-labels (see Ref. 15, p. 12). Indeed, without the assumption that there are such conservation "laws" (see Secs. 4.2.4 and 5.4.3), the proposed 2-space description of fundamental fermions would make no sense (e.g., flavors could *not* be distinguished). What are these conditions?

While flavor-defining global charge-labels should be conserved in *all* pair production or annihilation processes, once they are produced they are no longer required to be conserved in certain other processes. For example, above the GUT "unification"-energy where, according to $SU(5)$, leptons can be converted to quarks and vice versa,[4,5] these flavor-defining global charge-labels are certainly not conserved after pair production. At ultrahigh energies (e.g., the Planck energy) the "no-hair" theorem implies that flavors and flavor-defining global charge-labels lose their meaning altogether. Above this energy scale global flavor-defining charges are probably "eaten" or destroyed by virtual "wormholes" and/or "mini black-holes."[38]

Clearly, all of the 2-space global charge-labels defined in this book are considered to be conserved in all *conventional* "low"-energy flavor-preserving neutral-current processes (e.g., strong, electromagnetic or weak neutral-current processes). The charge labels $S+t = -\mathbf{U}^2$ or $S'+t' = -\mathbf{U}^2, Y, Y', B$ or $B' = L$ are considered to be conserved in all conventional *preferred* ("diagonal") flavor-changing weak processes. Finally, only the electric-charge labels and B or L are considered to be conserved in conventional ("off-diagonal") flavor-changing weak processes. These observations lead to the following "explanation" for the conventional "low"-energy flavor conservation-laws.

Since $S' + t' = -\mathbf{U}^2$ is supposed to be conserved in "diagonal" weak-processes involving leptons,[29] this is equivalent to saying that these processes conserve the attributes of "e^--ness," "μ^--ness" or "τ^--ness." Conventional, additively-conserved e, μ or τ quantum-numbers could be assigned to leptons, but their content would be no different than simply saying $S'+t'$ is conserved in these weak processes.[39,40]

Although it is not traditional to describe the "diagonal" weak-processes of quarks using the same kinds of quantum numbers as leptons (e.g., electron or muon numbers, etc.), the conservation of $S+t$ in these weak processes simply means that both the c and s quarks carry strangeness[35,36] (they are both strange quarks and $S+t = S$). Similarly, both the t and b quarks carry truth[35,36] (they are both truth-carrying quarks and $S+t = t$) and the u and d quarks carry neither strangeness nor truth ($S+t = 0$). According to the new scheme, conserved quantum numbers analogous to charm and beauty are carried by both quarks and leptons, so that a μ^- and a ν_μ, for example, differ by the "charm" $C' = +1$ quantum number (carried by the μ^- but not by the ν_μ), even though these particles both carry the same "strangeness" or "muonness" $S' = -1$ (see Table IV).[29]

How relevant the Planck energy is to elementary particle physics has not really been established, after all. It's merely a number with the dimensions of a mass that comes out of Newton's theory of gravity. —S. L. Glashow

All of physics, in my view, will be seen someday to follow the pattern of thermodynamics and statistical mechanics, of regularity based on chaos, of "law without law." —J. A. Wheeler

8. Why Does the Generalization $f \Rightarrow \mathbf{F}$ Describe Fundamental Fermions?

In this book I have explored some of the algebraic, geometric and physical consequences of a new *organizing principle* for fundamental fermions (see Sec. 1.2.6). The essence of the new organizing principle is the idea that the standard-model concept of scalar fermion-number f can be generalized. In particular, a "generalized fermion-number" consisting of a 2×2 real matrix $\mathbf{F}$ that acts on an *internal* 2-space, instead of *spacetime*, is taken to describe certain *internal* properties of fundamental fermions. This simple *matrix* generalization of the *number* f "automatically" introduces new degrees of freedom that "explain," among other things, family *replication* and the *number* of families. The resulting description epitomizes Occam's razor because it is both simple (e.g., f and $\mathbf{F}$ are square-roots of unity) and parsimonious—only the observed "low"-energy fermion (and antifermion) sector of the world can be accommodated.

The following list summarizes some of the more important characteristics of the new description of fundamental fermions:

- The abstract, internal 2-space that underlies the new description, together with certain associated global geometric-objects (scalars, vectors and matrices) are collectively taken to be "carried" in spacetime by *every* fundamental fermion.

- 2-vectors of the same type, which are carried by (describe) two different fundamental-fermions are *not* additive, only their associated 2-scalar "charges" are additive. Hence, the 2-spaces associated with two different particles are effectively *isolated* from one another.

- The new description "predicts" (*ex post facto*) that quark-lepton families are *replicated* and that there are only *three* families associated with light neutrinos. Taking this prediction seriously means that hypothetical families exhibiting conventional strong- and electroweak-interactions, which cannot be described by the 2-space geometry, simply do not exist in nature. For example, if additional families associated with *massive* neutrinos should exist, these families must be described in the same way (in the 2-space) as the three known families. That is, they must act like "excited" versions ("recurrences") of one of the three known families.

- The new 2-space description automatically introduces mutually-commuting (compatible) flavor-defining charge-like quantum numbers. These ("good") quantum numbers are "predicted" to be conserved in all *strong, electromagnetic* and *weak neutral-current interactions* (e.g., pair production) and many charge-changing weak interactions as well.

- Flavor eigenstates are represented by an appropriate combination of these mutually-commuting flavor-defining charges.

- The new description automatically links ("unifies") three familiar quantum dichotomies. These are the matter-antimatter, "up"-"down" and quark-lepton dichotomies. The three dichotomies are found to be essentially different *geometric* aspects of the 2-space.

- Effective "quantization" of the flavor-defining charges is achieved, in part, by "built-in" constraints placed on these charges by the 2-space non-Euclidean geometry.

- Two orthogonal coordinate-axes in the 2-space (these are one-dimensional invariant vector-spaces[30] associated with the Pauli reflection matrices $\pm\boldsymbol{\sigma}_z$) specify four *preferred* directions (positive and negative directions for each of two coordinate-axes). These are called "flavor directions" because they correspond to the four flavors (or antiflavors) per family (or antifamily), and because the electric charges of the fundamental fermions are "measured" along these directions *only*.

- The new description is more or less complementary to the description of fundamental fermions in the standard model of particle physics and GUT's such as $SU(5)$. That is, there are certain loose "connections," of an algebraic nature, between the local color-charges of $SU(5)$ and the new global flavor-defining 2-space charges.

- If nature is supersymmetric at some energy level, every family of spin $\frac{1}{2}$ *fundamental-fermions* (quarks and leptons) described by the 2-space will, by the definition of supersymmetry, be associated with a family of *fundamental-bosons* (squarks and sleptons) having essentially the "same" 2-space description as the fundamental fermions.

While the foregoing phenomenological description of fundamental fermions is simple and parsimonious, these qualities alone are insufficient to explain its apparent success. Something much more fundamental must be at work here.

8.1 Particle Physics at the Planck Level.

The fundamental explanation for elementary particles, and the fermion-sector of the world in particular, is probably to be found at the Planck level in a theory involving the *quantization* of gravity (e.g., superstrings). But, because of the uncertainties of Planck-level physics there is, as yet, no completely satisfactory theory of quantum gravity or of elementary particles. Peccei[41] has nicely summarized the theoretical situation in regard to the number of fermion families or generations N_g, and we will paraphrase and embellish some of his comments here.

Depending on the particular theory chosen, constraints on N_g vary considerably. In string theories having a large number of spatial-dimensions family-replication is the result of *compactification* of the "extra" space dimensions. The number of "handles" in the compactified-space is specified by a topological index called the Euler-characteristic,[14] which is supposed to equal twice the number of families. Unfortunately, there are many models involving compactification, but very few, if any, possess convincing mechanisms to limit the Euler-characteristic. Moreover, it may be possible to formulate string theories in 4-dimensional spacetime *without* the need of compactification. The result of all of this is that N_g is not yet calculated in these theories with any degree of certainty.

A wholly different type of explanation for N_g, which was proposed by H. B. Nielsen and collaborators,[13] involves the notion of *chaos* at the Planck level. These authors assume that dynamics is *random* at the Planck level and show that the effective Hamiltonian is not unique. This circumstance causes the "master group" or "world group" G to break down to a number of identical surviving replica-groups G_{surv}, each of which closely resembles the standard-model group. They further assume that the physics cannot distinguish between the surviving replica-groups and, therefore, that these groups must coexist with their respective "replica" fermion-representations. Because the resulting measurable gauge-fields are an average of the gauge fields from each surviving replica-group, a relation between the gauge-couplings and the number of families is established. Arguments along these lines seem to suggest that $N_g = 3$. But, there are many uncertainties and this result is not yet proven.

In spite of the uncertainties of Planck-level physics, and in spite of the variety of different theoretical approaches taken there, the following qualitative observations also seem to suggest that *some* form of Planck-level physics is responsible for the fermion-sector of the world and, in particular, for its phenomenological description via a "generalized fermion-number" **F**.

8.2 Planck-Level Physics and F.

It is well known that the "minimal unification" of *special* relativity and quantum mechanics (i.e., quantum field-theory) that was achieved by Dirac and others leads to *antiparticles* and *particle statistics*.[17-19] In particular, the concept of *scalar fermion-numbers* f emerges naturally, and unavoidably, from such a gravity-free unification (see Sec. 1.1). But, there is nothing in this minimal unification to suggest anything like the observed richness of the known elementary particles, in part because the scalar fermion-number f, by itself, is incapable of distinguishing different flavors of quarks and leptons. That is, fundamental fermions are "flavor-symmetric" at this level of unification.

By contrast to the foregoing minimal-unification, when *general* relativity and quantum mechanics are "unified," as in the superstring models[14] (i.e., when gravity is quantized), this global "flavor symmetry" is effectively broken, and a rich (too rich?) spectrum of elementary particles invariably emerges. Not only does the quantization of gravity lead to antiparticles and particle statistics as before (e.g., the scalar fermion-number f), but depending on the particular theory chosen (e.g., superstrings), a variety of different (distinguishable) types of "low"-energy elementary-fermions and bosons are an unavoidable consequence of the unification.

On very general grounds, then, the quantization of gravity ought to lead to a "low"-energy "flavor-asymmetrical" fermion sector that is described by something more *general* than f, possibly an internal description involving a *matrix* such as **F**. After all, there are numerous examples in nature where a relatively *asymmetrical* physical-situation is described by a *matrix* (e.g., **F**), whereas the corresponding *symmetrical* physical-situation is described by a *scalar* (e.g., f).

To cite a specific example, consider the dielectric-constant of matter[42] ϵ, which relates the applied electric-field vector **E** to the resultant electric-displacement vector $\mathbf{D} = \epsilon \mathbf{E}$. In a uniform, homogeneous and isotropic (spherically symmetric) liquid phase, ϵ can be described by a single *scalar* (or a scalar times the 3×3 identity matrix $\mathbf{I}_3$), whereas when the liquid undergoes a first-order phase-transition and "freezes" to form a less symmetrical crystalline phase, ϵ must be represented by a multicomponent 3×3 dielectric *matrix* ϵ. Moreover, the lower the degree of symmetry of the crystal resulting from this phase transition (e.g., triclinic crystals are less symmetric than cubic crystals), the greater the number of different scalar-components will be required for a complete specification of the associated dielectric matrix.

The foregoing very general observations, taken together, imply that something like the *matrix* generalization $f \Rightarrow \mathbf{F}$ *could*, in principle, provide an *internal* description of the observed "low"-energy flavor-asymmetrical "phase" of matter that presumably results from some type of unification of general relativity and quantum mechanics. In this picture, gravity effectively "breaks" the flavor-symmetry of fundamental-fermions that would hold in a world without gravity. And, this is more or less as it should be, since fermion *masses*, like all masses, are coupled to gravity and the appearance of mass (for whatever reasons) is essential if flavors are to be distinguished, i.e., the world is to be flavor-asymmetrical.

Perhaps some future theory of quantum gravity at the Planck level will automatically show why **F** provides an acceptable "macroscopic" or "low"-energy internal description of the presently known fermions. Meanwhile, many questions remain.

Assuming that the new 2-space description is a valid "macroscopic" or "low"-energy internal description of fundamental fermions, what, if any, are its further consequences? Can we use this 2-space to make predictions of new phenomena beyond those we have touched on here? Is there a legitimate "macroscopic" or "low"-energy *quantum dynamics* underlying the 2-space description of such things as the "evolution" of vector triads (see Sec. 5.4.3)?

Connections between the 2-space, and internal symmetries based on $SU(3)_c$ and $SU(5)$, have been alluded to throughout this book. Is it possible

8.2. PLANCK-LEVEL PHYSICS AND F

to develop a comprehensive "macroscopic" or "low"-energy quantum theory (i.e., an "extension" of the standard model) that places these guessed-at connections on a sound theoretical basis? Is there a connection between the family-hierarchy associated with the matrix **R** (see Sec. 6) and that associated with the average particle-mass (of a family)? Is there any connection between **C**-invariant geometric objects in the 2-space (e.g., see the matrix **A** in Appendix B) and **C**-invariant particle masses? What are the full theoretical implications of describing the matter-antimatter, "up"-"down" and quark-lepton dichotomies as different *geometric* aspects of the *same* thing, namely, the 2-space?

In closing I would like to raise several additional speculative, but hopefully thought-provoking, questions. First, it is a fact that both the internal 2-space and 4-dimensional spacetime have "Lorentzian" metrics. And, in both cases, but apparently for very different reasons, this is necessary to obtain a self-consistent description of antimatter [e.g., see Ref. 21 (including note) and compare Eqs. (1.1) and (1.2) with Eqs. (1.12) and (1.13), respectively]. Is this "accidental" or is it an indication of some deeper connection between these otherwise very different spaces, and if so, what is the connection? Second, because M_c counts the number of strong-color *configurations* for a fundamental fermion having a given mass-energy and electric charge, the dimensionless group-parameter appearing in (1.24) and given by (5.23), namely, $v = \ln M_c$, superficially resembles a dimensionless *entropy* $(S/k = v)$. Is this also an "accident" having to do, in this case, with the choice of variable (v versus M_c), or is this an indication of some deeper connection with "random" physical processes, perhaps at the Planck level? Third, are the two previous questions just different aspects of the *same* question? Fourth, and finally, conventional strong-interactions seem to constrain M_c to be 1 or 3 only. But, the only constraint placed on M_c (and the group-parameter $v = \ln M_c$) by the 2-space, and the spin-statistics theorem, is that M_c must be a positive odd-integer (see Sec. 5.2.1). Is it possible that exotic (massive) "quarks" characterized by $M_c > 3$ exist? If $M_c = 5$, these hypothetical "quarks" would carry electric charges $\pm 3/5$, $\mp 2/5$ and "baryon" numbers $\pm 1/5$. Then, 5 such "quarks" would be bound by exotic $SU(5)$-based "strong"-color forces to form a composite "baryon."

Probably, this sort of thing does not happen. But, keep in mind that current laboratory experiments probe energies some 16 orders of magnitude *below* the Planck energy. This huge gap leaves ample room for surprise.

Answers to the many questions raised, but left unanswered, in this book will have to await further theoretical and experimental developments.

Appendix A. Three Quantum "Dichotomies" and $SU(5)$.

Consider first-family fundamental fermions (u, d, e^-, ν_e) and their associated antiparticles $(\bar{d}, \bar{u}, \bar{\nu}_e, e^+)$. In $SU(5)$ the internal strong- and weak-color content of any left-handed (L)- or right-handed (R)-type fundamental-fermion, and its net strong-color, can be represented using the following symbolic form:[5]

$$(\text{flavor symbol})_{L \text{ or } R} (R, W, B, G, Y)_{(\text{net strong color})}. \tag{A1}$$

The color symbols within the second pair of parentheses from the left are the three strong-color charges, namely, the "red" R, "white" W and "blue" B, color charges and the two weak-color charges, namely, the "green" G and "yellow" Y color charges. All color charges are dichotomous, i.e., they have just two values, namely, $+1$ or -1 (sometimes 0 and 1 are also used). Hence, in $SU(5)$ *the five color-values (R, W, B, G, Y) are analogous to a "five-bit word."*[5]

The electric charges Q carried by any one of the L- or R-type fundamental-fermions or antifermions can be expressed by[4,5]

$$Q = G/2 + Y(\text{weak})/2, \tag{A2}$$

where the weak-hypercharge is given by

$$Y(\text{weak}) = -1/3 \, (R + W + B). \tag{A3}$$

Note that $Y(\text{weak})$ has the same value for the "up"- and "down"-type flavors in any given flavor doublet. In fact, $Y(\text{weak})$ for L-type fundamental-fermions or R-type fundamental-antifermions, can also be expressed in terms of the *baryon number B*, or *lepton number L*, by the simple relation

$$Y(\text{weak}) = B \text{ or } -L, \tag{A4}$$

where B and L are not to be confused with the strong-color blue or the subscript for left-handed fermions, respectively.

In $SU(5)$ the fact that there are only "up"-"down"-type flavors or flavor-doublets (the "up"-"down"-type flavor dichotomy) can be "explained" (described) by noting the color content of corresponding "up"-"down"-type pairs, i.e.,

$$\left.\begin{array}{l} u_L(+,-,-,+,-)_{\text{red}} \leftrightarrow d_L(+,-,-,-,+)_{\text{red}} \\ u_L(-,+,-,+,-)_{\text{white}} \leftrightarrow d_L(-,+,-,-,+)_{\text{white}} \\ u_L(-,-,+,+,-)_{\text{blue}} \leftrightarrow d_L(-,-,+,-,+)_{\text{blue}} \\ \nu_{eL}(+,+,+,+,-)_0 \leftrightarrow e_L^-(+,+,+,-,+)_0 \end{array}\right\} \quad (A5)$$

Clearly, left-handed "up"-type fundamental-fermions are converted to left-handed "down"-type fundamental-fermions, and vice versa, by changing the signs of both the G and Y color charges [see A(1)]. If handedness is ignored, these "up"-"down"-type flavor-conversions are accomplished by simply changing the sign of the G color-charge.[4,5] Hence, in $SU(5)$ and/or $SU(2)_L$, the dichotomous choice $G = \pm 1$ may be said to "explain" the "up"-"down"-type flavor dichotomy.

Although $SU(5)$ "explains" (describes) the observed flavor-doubling it provides no explanation, whatever, for the "family problem," i.e., for the fact that there exist in nature six similar flavor-doublets together with their antiparticle counterparts. Therefore, an entirely different kind of underlying "two-dimensional" structure (e.g., the new 2-space), which is responsible for the observed "up"-"down"-type flavor-dichotomy of fundamental fermions, cannot be ruled out by $SU(5)$.

In $SU(5)$ the matter-antimatter dichotomy is "explained" (described) by the fact that each of the five color-charges has a **C**-reversed counterpart of opposite sign. Changing the sign of all five color-charges changes L-type particles (antiparticles) to R-type antiparticles (particles) and vice versa, i.e.,

$$\left.\begin{array}{l} u_L(+,-,-,+,-)_{\text{red}} \leftrightarrow \bar{u}_R(-,+,+,-,+)_{\text{antired}} \\ u_L(-,+,-,+,-)_{\text{white}} \leftrightarrow \bar{u}_R(+,-,+,-,+)_{\text{antiwhite}} \\ u_L(-,-,+,+,-)_{\text{blue}} \leftrightarrow \bar{u}_R(+,+,-,-,+)_{\text{antiblue}} \\ d_L(+,-,-,-,+)_{\text{red}} \leftrightarrow \bar{d}_R(-,+,+,+,-)_{\text{antired}} \\ d_L(-,+,-,-,+)_{\text{white}} \leftrightarrow \bar{d}_R(+,-,+,+,-)_{\text{antiwhite}} \\ d_L(-,-,+,-,+)_{\text{blue}} \leftrightarrow \bar{d}_R(+,+,-,+,-)_{\text{antiblue}} \\ \nu_{eL}(+,+,+,+,-)_0 \leftrightarrow \bar{\nu}_{eR}(-,-,-,-,+)_0 \\ e_L^-(+,+,+,-,+)_0 \leftrightarrow e_R^+(-,-,-,+,-)_0 \end{array}\right\} \quad (A6)$$

If handedness is ignored, any flavor can be converted to its antiflavor counterpart by changing the signs of only the R, W, B and G color-charges.

Finally, in $SU(5)$ and/or $SU(3)_c$, the quark-lepton "dichotomy," or more accurately, the "strong-electroweak" dichotomy, comes about because quarks are strong-color triplets (quarks possess nonzero net strong-color), whereas, leptons are strong-color singlets (leptons possess zero net strong-color). Notice that the following quark-lepton or lepton-quark "transformations," namely,

$$\left. \begin{array}{r} \left. \begin{array}{r} u_L(+,-,-,+,-)_{\text{red}} \\ u_L(-,+,-,+,-)_{\text{white}} \\ u_L(-,-,+,+,-)_{\text{blue}} \end{array} \right\} \leftrightarrow e_L^-(+,+,+,-,+)_0 \\ \\ \left. \begin{array}{r} d_L(+,-,-,-,+)_{\text{red}} \\ d_L(-,+,-,-,+)_{\text{white}} \\ d_L(-,-,+,-,+)_{\text{blue}} \end{array} \right\} \leftrightarrow \nu_{eL}(+,+,+,+,-)_0 \end{array} \right\}, \qquad (A7)$$

always require two of the like-signed strong-color charges (i.e., R, W or B) to be changed and "up" ("down")-type quarks are always changed in the process to "down" ("up")-type leptons, regardless of their handedness.

Appendix B. The Matter-Antimatter "Reflection".

The transformation $-\boldsymbol{\sigma}_x$, which generates a realization of the discrete group Z_2, namely, $Z_2(-\boldsymbol{\sigma}_x) = \{-\boldsymbol{\sigma}_x, \mathbf{I}_2\}$, is analogous to *charge-conjugation* $\mathbf{C}$ in its action on the global charge-labels defined on the internal 2-space, i.e., it changes their signs. In general, any real charge-vector $\mathbf{W}$, expressed as a column matrix, is transformed into its antiparticle counterpart $\mathbf{W}^c$ by the rule

$$-\boldsymbol{\sigma}_x \mathbf{W} = \mathbf{W}^c \neq -\mathbf{W}, \tag{B1}$$

where the superscript c here serves only to designate antimatter and does *not* mean complex conjugation. Thus, $-\boldsymbol{\sigma}_x$ converts a quark electric-charge vector such as $\mathbf{Q}_q = \mathbf{Q}_q(v, f)$ in (5.4) to its orthogonal antiparticle-counterpart $\mathbf{Q}_q^c = \mathbf{Q}_q(v, -f)$, namely,

$$-\boldsymbol{\sigma}_x \mathbf{Q}_q(v, f) = \mathbf{Q}_q(v, -f), \tag{B2}$$

where

$$q_1(v, f) = -q_2(v, -f) \tag{B3}$$

and

$$q_2(v, f) = -q_1(v, -f). \tag{B4}$$

Clearly, $-\boldsymbol{\sigma}_x$ changes the sign of f. That is, it changes the signs of the baryon number (5.9) and the component electric-charge labels (B3) and (B4). In fact, $-\boldsymbol{\sigma}_x$ acts on all of the nonzero (global) charge labels carried by each of two quark, or lepton, flavors (in any given quark, or lepton, flavor doublet) to convert them to their antiparticle counterparts.

This is easily demonstrated by noting that $-\boldsymbol{\sigma}_x$ converts every column vector having the form $\{x_1, x_2\}$, in the expression $\mathbf{Q} = \mathbf{U} + \mathbf{V}$ representing a flavor doublet, to a column vector having scalar-components of opposite sign and reversed order, namely, $-\boldsymbol{\sigma}_x\{x_1, x_2\} = \{-x_2, -x_1\}$. Therefore, $-\boldsymbol{\sigma}_x$ also changes the sign of any associated scalar quadratic form such as

$$(x_1, x_2)\{x_1, x_2\} = x_1^2 - x_2^2 \tag{B5}$$

83

or

$$(x_1, x_2)\{y_1, y_2\} = x_1 y_1 - x_2 y_2. \tag{B6}$$

While $-\boldsymbol{\sigma}_x$ is analogous to charge-conjugation $\mathbf{C}$, it is *not* identical to it for several reasons. First, $\mathbf{C}$ is a 4×4 matrix that operates on spinors (in the case of fermions), whereas, $-\boldsymbol{\sigma}_x$ is a 2×2 matrix that acts strictly on real 2-vectors. Second, two applications of $\mathbf{C}$ introduces an unobservable 180 degree phase shift,[24] i.e., $\mathbf{C}^2 = -\mathbf{I}_4$, where $\mathbf{I}_4$ is a 4×4 identity matrix. By contrast, two applications of $-\boldsymbol{\sigma}_x$ yields the 2×2 identity matrix $\mathbf{I}_2$. Third, $-\boldsymbol{\sigma}_x$ only affects the $\mathbf{C}$-reversing global charge-labels. Appropriately, it has no direct effect on the five local (gauge) color-charges R, W, B, G and Y. But, *to maintain a self-consistent physical picture, it must be imagined that when* $-\boldsymbol{\sigma}_x$ *acts to change the signs of electric charges, an appropriate* $\mathbf{C}$ *must also simultaneously act on appropriate spinor-type wave functions to change the signs of the internal color-charges.*[28]

A 2×2 square matrix $\mathbf{M}$ is transformed into its $\mathbf{C}$-reversed counterpart by a similarity transformation

$$(-\boldsymbol{\sigma}_x) \mathbf{M} (-\boldsymbol{\sigma}_x) = \mathbf{M}^c. \tag{B7}$$

For example, the matrix $\mathbf{F}(v)$, which represents the global "generalized fermion-number," transforms under $-\boldsymbol{\sigma}_x$ in a charge-like manner, i.e., it is obviously "$\mathbf{C}$-reversing" since

$$(-\boldsymbol{\sigma}_x) \mathbf{F}(v) (-\boldsymbol{\sigma}_x) = \mathbf{F}^c(v) = -\mathbf{F}(v). \tag{B8}$$

Although it is less obvious, there also exist matrices such as $\mathbf{R}$ (see Sec. 6.3), which are $\mathbf{C}$-reversing even though $\mathbf{R}^c \neq -\mathbf{R}$. Finally, there are symmetric matrices [e.g., the matrix $-\boldsymbol{\sigma}_x$ or matrices $(\mathbf{R} + \mathbf{R}^c)^p$, where p is an integer] of the form

$$\mathbf{A} = \begin{pmatrix} a & b \\ b & a \end{pmatrix} \tag{B9}$$

that are "$\mathbf{C}$-invariant" since

$$(-\boldsymbol{\sigma}_x)(\mathbf{A})(-\boldsymbol{\sigma}_x) = \mathbf{A}. \tag{B10}$$

Appendix C. Identifying the Symmetry Class of Flavor Doublets.

As indicated in Sec. 5.4, if a charge-vector triad is not in symmetry class $S(I)$, it must be in symmetry class $S(II)$. The purpose of this appendix is to determine the symmetry classes of the various flavor doublets. We will show that only the four first-family flavor-doublets (u, d), $(\bar{d}, \bar{u})$, (e^-, ν_e) and $(\bar{\nu}_e, e^+)$ fall in $S(I)$, which means that all other flavor doublets, corresponding to all other families, necessarily, fall in $S(II)$. The proposed $S(I)$ symmetry classification of the charge-vector triad ($\mathbf{Q}_q$ or $\mathbf{Q}_\ell$, $\mathbf{U}$, $\mathbf{V}$), representing the global charge-label content of the first-family flavor-doublets, will now be described in detail.

Because the foregoing (four) first-family flavor doublets are all associated with $\mathbf{U}$-vectors that are eigenvectors of $-\boldsymbol{\sigma}_x$, they are all in the same $S(I)$ symmetry class. Hence, the argument need only be presented for the (u, d) and $(\bar{d}, \bar{u})$ quark flavor-doublets. This will lead to the derivation of the third-component of global-isospin T_3 and the global-hypercharge Y, for the first-family flavor-eigenstates represented by the ket-vectors $|T_3, Y\rangle$.

The internal (global) charge-label T_3, unlike the (global) electric-charge labels q_1, q_2 or baryon number B, which apply to both $S(I)$- and $S(II)$-type quark flavor-doublets, will be the direct consequence of the $S(I)$ symmetry of the charge-vector triad representing (u, d). Thus an alternate way to derive the $\mathbf{U}$-vector for (u, d) will have been found. See Sec. 4.2.1 for the other way to derive this particular $\mathbf{U}$-vector.

In $S(I)$, $\mathbf{F}(v)$, which is given by (5.1) is, by definition, supposed to act once on the vector-triad

$$\mathbf{Q}_q(v, f) = \mathbf{U} + \mathbf{V}, \tag{C1}$$

giving back the *same* flavor-doublet-specific charge-vectors $\mathbf{U}$ and $\mathbf{V}$. Using (5.2) and $f = \pm 1$ or $f = 1/f$; when $\mathbf{F}(v)$ is applied once to (C1) the result is

$$\mathbf{Q}_q(v, f) = f\mathbf{F}(v)\mathbf{U} + f\mathbf{F}(v)\mathbf{V}. \tag{C2}$$

Therefore, the $S(I)$-type congruence, in which (C1) is congruent to (C2), can only be achieved if

$$\mathbf{F}(v)\mathbf{V}'' = \mathbf{U}, \tag{C3}$$

or, equivalently,

$$\mathbf{F}(v)\mathbf{U} = \mathbf{V}'', \tag{C4}$$

where $f\mathbf{V} = \mathbf{V}''$. Note that Eqs. (C3) and (C4) are required, because $\mathbf{Q}_q(v, f)$ is, by definition, an eigenvector of $\mathbf{F}(v)$, but in general $\mathbf{U}$ and $\mathbf{V}$ are not. Using these results and definitions, (C2) becomes

$$\mathbf{Q}_q(v, f) = \mathbf{U} + f\mathbf{V}'', \tag{C5}$$

where $+f$ goes to $-f$ when going from matter to antimatter. Clearly, the special charge-vector triad represented by (C5) is, by construction, congruent to itself after one application of $\mathbf{F}(v)$.

In other words, the charge-vectors $\mathbf{U}$ and $\mathbf{V}''$ are required to be independent of the fermion number f (i.e., they are the same for quarks and antiquarks) to establish an $S(I)$-type congruence. In particular, applying the matter-antimatter reflection $-\boldsymbol{\sigma}_x$ to (C5), the result for antiquarks must be ($+f$ goes to $-f$ under $-\boldsymbol{\sigma}_x$)

$$-\boldsymbol{\sigma}_x\mathbf{U} - f\boldsymbol{\sigma}_x\mathbf{V}'' = \mathbf{U} - f\mathbf{V}''. \tag{C6}$$

This last result can only mean that $\mathbf{U}$ and $\mathbf{V}''$ are both *eigenvectors* of the known matter-antimatter reflection $-\boldsymbol{\sigma}_x$. That is,

$$-\boldsymbol{\sigma}_x\mathbf{U} = \mathbf{U} \tag{C7}$$

and

$$-\boldsymbol{\sigma}_x\mathbf{V}'' = -\mathbf{V}''. \tag{C8}$$

The reader should note that the derivation of (C7) and, therefore, $\mathbf{U}$ was arrived at here independently of the requirements set forth in Sec. 4.2.1.

Given $\mathbf{F}(v)$ or $\mathbf{Q}_q$, and using Eqs. (C7), (C8) and (3.1), $\mathbf{U}$ and $\mathbf{V}''$ can easily be derived and the global charge-label components of these vectors can simultaneously be given a simple interpretation, namely (see also Tables III and V),

$$\mathbf{U} = \begin{pmatrix} +1/2 \\ -1/2 \end{pmatrix} = \begin{pmatrix} T_3^+ \\ T_3^- \end{pmatrix} \tag{C9}$$

and

$$\mathbf{V}'' = \begin{pmatrix} 1/6 \\ 1/6 \end{pmatrix} = f \begin{pmatrix} Y/2 \\ Y/2 \end{pmatrix}, \tag{C10}$$

or $\mathbf{V} = f\{1/6, 1/6\}$ for both the first-family (u,d) quarks with $f = +1$ and the $(\bar{d}, \bar{u})$ antiquarks with $f = -1$. Thus, the ket-vectors (in abbreviated form) describing the u and d quark flavors are found to be $|T_3^+, Y\rangle$ and $|T_3^-, Y\rangle$, respectively.

In summary, the description of the (u,d) and $(\bar{d}, \bar{u})$ quark flavor-doublets via the discrete, global $Z_2(v)$ symmetry-group realization, and the symmetry class $S(I)$, has resulted in the derivation of a quantity that acts like the third-component of global isospin T_3. Because leptons have been associated with inverted $\mathbf{U}$-vectors (see Sec. 2.5.3), the first-family flavor-doublets (e^-, ν_e) and $(\bar{\nu}_e, e^+)$ also fall in $S(I)$. Hence, all remaining flavor-doublets in "higher" families must fall in $S(II)$. In particular, the flavor doublets associated with the second family (c, s, μ^-, ν_μ) or $(\bar{s}, \bar{c}, \bar{\nu}_\mu, \mu^+)$ fall in what we call $S(II_1)$ or $S(II_2)$, respectively, while the flavor doublets associated with the third family (t, b, τ^-, ν_τ) or $(\bar{b}, \bar{t}, \bar{\nu}_t, \tau^+)$ fall in what we call $S(II_2)$ or $S(II_1)$, respectively. Here, the index 1 (2) refers to the fact that the associated $\mathbf{U}$-vector lies in the one-dimensional invariant vector-space associated with $+\boldsymbol{\sigma}_z$ ($-\boldsymbol{\sigma}_z$), namely, the X_1 (X_2) axis. See Tables III through VI in the main text for further details.

Footnotes and References

1. P. J. Kernan and L. M. Krauss, *Phys. Rev. Lett.*, 72, 3309 (1994).

2. G. S. Abrams, et al., *Phys. Rev. Lett.*, 63, 2173 (1989).

3. CDF Collaboration, F. Abe et al., *Phys. Rev.* **D50**, 2966 (1994); *Phys. Rev. Lett.* **73**, 224 (1994).

4. C. Quigg, *Gauge Theories of the Strong, Weak, and Electromagnetic Interactions*, (The Benjamin/Cummings Publishing Co., Reading, Mass., 1983). In the standard model based on the spontaneously-broken local gauge-group $SU(3)_c \times SU(2)_L \times U(1)_Y$, left (right)-handed fundamental-fermions are doublets (singlets) under the weak-isospin group $SU(2)_L$, while the relative phases of the particles under the gauge-group $U(1)_Y$ are fixed by definite values of weak-hypercharge Y(weak). The gauge group $SU(3)$ color or $SU(3)_c$ represents an unbroken symmetry in which quarks (leptons) are strong-color triplets (singlets), while the $SU(2)_L \times U(1)_Y$ symmetry is spontaneously-broken, giving quarks, leptons and intermediate vector-bosons mass. In a local five-color grand unified theory or GUT such as $SU(5)$, the fundamental fermions are assumed to carry combinations of five different local color-charges. These are the three strong-colors of $SU(3)_c$ and quantum chromodynamics (QCD), namely, "red" R, "white" W and "blue" B, and two weak-colors, namely, "green" G, which defines weak-isospin T_3(weak) $= \frac{1}{2}G$ [for L-type fermions and R-type antifermions only; T_3(weak) $= 0$ for L-type antifermions and R-type fermions] and "yellow" Y. Each of these C-reversing charges can be represented by ± 1, and the allowed combinations of five color-charges are always consistent with strong-color triplets (singlets) for quarks (leptons). In $SU(5)$, there are $2^5 = 32$, L- and R-type fundamental-fermions and antifermions (assuming all neutrinos have small masses) corresponding to the four

flavors and antiflavors per family. These consist of two L- and two R-type leptons (antileptons) and six L- and six R-type quarks (antiquarks). Only the four local-colors (R, W, B and G) make contributions to the total electric-charge label, while the fifth weak-color Y (together with G), determines the helicity of the fermion. The weak-hypercharge Y(weak) is a linear combination of contributions from the three strong-colors, namely, $Y(\text{weak}) = -\frac{1}{3}(R + W + B)$. Hence, the local electric-charge of any L- or R-type fundamental-fermion or antifermion in $SU(5)$ is given by $Q(\text{local}) = G/2 + Y(\text{weak})/2$. $SU(5)$ has several possible fermion representations, namely, two irreducible representations; the **15** and **45** and a reducible representation $\mathbf{5^* \oplus 10}$. But, only the $\mathbf{5^* \oplus 10}$ (wherein multiplets contain both fermions and antifermions, leptons and quarks) representation leads to the observed spin $\frac{1}{2}$ fundamental-fermion family structure. In $SU(5)$ all L-type quark or lepton flavor-doublets, each consisting of "up"- and "down"-type flavors ($u-d, c-s, t-b, \nu_e-e^-, \nu_\mu-\mu^-, \nu_\tau-\tau^-$), are weak-isospin doublets with $T_3(\text{weak}) = \pm\frac{1}{2}$, while the associated weak-hypercharge values are $Y(\text{weak}) = \frac{1}{3}$ for all quarks and $Y(\text{weak}) = -1$ for all leptons. Besides gluons g, which transform only strong-color charges, and various weak intermediate vector-bosons ($Z^0, W^\pm$), which transform only weak-color charges (the photon γ has no effect on color charge), various other "gluons" (X and Y) are assumed to exist in a GUT such as $SU(5)$. With the full complement of "gluons," the five color-charges can be transformed among each other in all possible ways. For example, these color-transformations mediate not only the well-known strong- and electroweak-interactions, but also interactions that convert fermions to antifermions and quarks to leptons.

5. F. Wilczek and B. Devine, *Longing For The Harmonies*, (W. W. Norton and Company, 1988), p. 255.

6. U. Amaldi, W. de Boer and H. Furstenau, *Phys. Lett.* B, 260, 447 (1991).

7. S. Dimopoulos, S. A. Raby and F. Wilczek, *Phys. Today*, 25 (October 1991).

8. H. Arason, D. J. Castano, et al., *Phys. Rev. Lett.*, **67**, 2933 (1991).

9. O. Nachtmann, *Elementary Particle Physics—Concepts and Phenomena*, (Springer-Verlag, Berlin, 1990).

10. T. Goldman and M. M. Nieto, *Los Alamos Science*, 11, 114 (1984).

11. H. Georgi, *Seventh Workshop on Grand Unification/ICOBAN86*, Jiro Arafune, Ed., (World Scientific 1987), pp. 303–337.

12. t'Hooft, G., Recent Developments in Gauge Theories. In *Cargese Lectures 1979*, G. t'Hooft et al., Eds., (New York: Plenum Press, 1980).

13. Nielsen, H. B., Gauge Theories of the Eighties, *Proceedings of the Arctic School of Physics*, R. Raito and J. Lingfors, Eds. (Berlin: Springer, 1983) and by Nielsen, H. B., D. L. Bennett and N. Brene, in *Developments in Quantum Field Theory*, Niels Bohr Centennial Conference 1985 (Amsterdam: North Holland, 1985).

14. M. B. Green, J. H. Schwartz and E. Witten, *Superstring Theory, Vol. 2*, (Cambridge University Press, New York, 1987), p. 408.

15. J. Bernstein, *Elementary Particles and Their Currents*, (W. H. Freeman and Co., San Francisco, 1968).

16. J. D. Bjorken and S. D. Drell, *Relativistic Quantum Fields*, (McGraw-Hill Book Company, New York, 1965), p. 60. In a relativistically-invariant (Lorentz invariant) quantum field-theory involving creation and destruction operators, it is both natural and mandatory to assign global, additively-conserved, fermion numbers to *all* particles obeying Fermi statistics ($f = f_m = +1$ for fermions and $f = f_a = -1$ for antifermions), whether or not they are composite particles.

17. R. P. Feynman and S. Weinberg, *Elementary Particles And The Laws of Physics—The 1986 Dirac Memorial Lectures*, (Cambridge University Press, Cambridge, 1987), pp. 1–59.

18. R. P. Feynman, *The Theory of Fundamental Processes*, (W. A. Benjamin Inc., Reading, MA., 1962), pp. 25–28.

19. Dirac, P. A. M., *Proc. R. Soc. London*, *117*, 610, *118*, 351 (1928).

20. S. Perlis, *Theory of Matrices*, (Dover Publications Inc., New York, 1991).

21. M. Hammermesh, *Group Theory And Its Applications To Physical Problems*, (Addison-Wesley, Reading, MA., 1962), p. 84. Why two dimensions? Regardless of the dimension of the representation of **F** (i.e., the dimension of the space on which **F** "acts") the eigenvalues of **F** ought to involve some combination of the two scalar fermion-numbers $f_m = +1$ and $f_a = -1$. For example, the simplest representations of **F** are just the two distinct one-dimensional representations, namely, $f_m = +1$ and $f_a = -1$. Some care must be taken when speaking of "dimensions" here. A scalar in a D-dimensional space always has $N = D^r = D^0 = 1$ component. Although the space is D-dimensional, this one-component object is sometimes referred to as a "one-dimensional" representation. Evidently, all higher-dimensional representations of **F** can be constructed by first forming the *direct sum* of a number n of these one-dimensional representations, which equals the desired dimension. Then, the particular representation of **F** will be a *matrix* that is equal to this direct sum, up to some *similarity transformation*. We simply *guess* that because particles and antiparticles always appear together in pair-production processes, the number of times the one-dimensional representation $f_m = +1$ appears in any such direct-sum representing a higher-dimensional representation of **F**, ought to be the same as the number of times the one-dimensional representation $f_a = -1$ appears in this same representation. Thus, it is "reasonable" to assume that a physically acceptable representation of **F** must be an *even-dimensional representation*, which is equal, up to an appropriate similarity transformation, to the direct sum of any integral number of distinct one-dimensional representations of **F**. The simplest choice, which does not involve a repetition or "degeneracy" in f, is one for which **F** *is a two-dimensional representation* similar to a 2×2 matrix consisting only of the nonzero elements $f_m = +1$ and $f_a = -1$ on its diagonal, and zero off-diagonal elements.

22. A. W. Joshi, *Elements of Group Theory for Physicists*, (Wiley Eastern Limited, New Delhi, 1983).

23. L. Rosenfeld, *Nuclear Forces*, (North-Holland Publishing Co., Amsterdam Intersource Publishing, Inc., New York 1948), p. 43.

24. C. Itzykson and J-B. Zuber, *Quantum Field Theory*, International Series in Pure and Applied Physics, (McGraw-Hill, New York, 1980), p.

693. Applying charge conjugation $\mathbf{C} = i\gamma^2\,\gamma^0$ (γ^μ are the 4×4 Dirac γ matrices) twice to a fermion spinor introduces an unobservable 180 degree phase-factor since $\mathbf{C}^2 = -\mathbf{I}_4$. But, since $\mathbf{X}$ acts on *real* 2-vectors instead of *spinors*, two applications of $\mathbf{X}$ should be equivalent to the *identity*, i.e., $\mathbf{X}^2 = \mathbf{I}_2$ or $\mathbf{X} = \mathbf{X}^{-1}$.

25. P. A. M. Dirac, *The Principles of Quantum Mechanics* (Fourth Edition revised), (Oxford University Press 1976).

26. Y. Ne'eman, *Algebraic Theory of Particle Physics*, (W. A. Benjamin, Inc., New York, 1967), p. 13; H. Georgi, *Lie Algebras in Particle Physics*, (The Benjamin-Cummings Publishing Co., New York, 1982), p. 52. If only strong- and electromagnetic-interactions are considered, the three global quark flavor-eigenstates u, d and s can be represented in global $SU(3)$-flavor by ket-vectors of the form $|Q(\text{global}), T_3, B, S\rangle$ or $|T_3, Y\rangle$, where T_3 is the third-component of global isospin, B is the baryon number, S is the strangeness, $Y = B + S$ is the global hypercharge and $Q(\text{global}) = T_3 + Y/2$ is the electric charge of the flavor eigenstate in question. Clearly, both T_3 and Y are global color-independent charge-like eigenvalues associated with a mutually-commuting subset of $SU(3)$-flavor generators, i.e., a Cartan subalgebra of $SU(3)$-flavor. These global charge-labels are conserved ("good" quantum numbers) in all strong- and electromagnetic-interactions involving these flavor eigenstates. Unfortunately, global $SU(N)$-flavor symmetries for $N > 3$ are badly broken by quark mass terms. Moreover, such symmetries do not apply to the electroweakly-interacting leptons.

27. M. Gell-Mann and Y. Ne'eman, *The Eightfold Way*, (W. A. Benjamin, Inc., New York, 1964).

28. Strictly speaking, besides the specification of global charges, the overall quantum state of a fundamental fermion would, necessarily, involve a specification of the spin state, the energy-momentum state and so on, together with a specification of the particular mix of local color (gauge)-charges R, W, B, G and Y carried by each fundamental fermion. This color-mix would be determined, in turn, by the complementary, local $SU(5)$ color-dependent gauge description.

29. The 2-space provides a method for introducing a variety of global flavor-defining quantum numbers for both quarks and leptons. In general, these quantum numbers are different than the weak-hypercharge or weak-isospin of $SU(5)$, which define flavors in the terms of internal color-charges or functions of these charges. The new flavor-defining quantum numbers for quarks (unprimed quantities) closely resemble conventional global quantum-number assignments (e.g., strangeness, charm, third-component of isospin, etc.). But, the new flavor-defining quantum numbers for leptons (primed quantities) are unconventional.

30. A. J. Pettofrezzo, *Matrices and Transformations*, (Dover Publications Inc., New York, 1966), p. 87–89.

31. We are proceeding (for the time-being only) as if families other than the first family did not exist. But, it is clear that $q_1 = +\frac{2}{3}$ ($q_2 = -\frac{1}{3}$) is the electric charge of the u, c or $t(d, s$ or $b)$ quark carrying $B = +\frac{1}{3}$, and $q'_1 = -1(q'_2 = 0)$ is the electric charge of the e^-, μ^- or τ^- (ν_e, ν_μ or ν_τ) lepton carrying $L = +1$. For the **C**-reversed antifermion families, e.g., $(\bar{d}, \bar{u}, \bar{\nu}_e, e^+)$, $(\bar{s}, \bar{c}, \bar{\nu}_\mu, \mu^+)$ and $(\bar{b}, \bar{t}, \bar{\nu}_\tau, \tau^+)$ we have B, L and electric-charge values reversed in sign.

32. These **U**-vectors, all of which lie along the X_1 or X_2 internal coordinate-axes (i.e., they actually reside in the one-dimensional invariant vector-spaces associated with $\pm\boldsymbol{\sigma}_z$), are said to "reside" in the *physical* quadrants II and IV rather than in the *unphysical* quadrants I and III even though, strictly speaking, they reside in *neither* quadrant.

33. In the language of group theory, $\mathbf{F}(v)$ is a *generator* of a two-element subgroup $Z_2(\mathbf{F}) = \{\mathbf{F}, \mathbf{I}_2\}$, of the direct-product group $SO(1,1) \otimes Z_2(\boldsymbol{\sigma}_z)$ or $SO(1,1) \otimes \{\boldsymbol{\sigma}_z, \mathbf{I}_2\}$.

34. J. Rosen, *A Symmetry Primer for Scientists*, (John Wiley and Sons, New York, 1983).

35. The t and b global charge-label assignments for quarks dictated by the new description are unconventional but apparently harmless. Usually, only the t quark is assigned a *truth* or *top* charge-label $t = 1$, but here both the t and the b quarks carry this charge label just as the c and s quarks both carry *strangeness* $S = -1$ in this scheme. The b quark is conventionally assigned a *beauty* or *bottom* charge label $b = +1$, but here it is assigned $b = -1$.

36. K. Huang, *Quarks, Leptons and Gauge Fields*, (World Scientific Publishing Co., Singapore, 1982), p. 45. Note that the c and s quarks properly carry the *same* strangeness. Huang points out that the GIM mechanism, together with the $\Delta S = \Delta Q$ selection-rule, requires that the charmed quark c carry the same strangeness as the strange quark s, namely, $S = -1$.

37. S. L. Glashow, J. Iliopoulos and L. Maiani, *Phys. Rev. D*, **2**, 1285 (1970).

38. C. W. Misner, K. S. Thorne and J. A. Wheeler, *Gravitation*, (W. H. Freeman and Company, San Francisco, 1973), p. 1214; R. Holman, et al., *Phys. Rev. Lett.*, **69**, 1489 (1992).

39. Since the weak-interaction flavor, or mass, eigenstates generally don't mix, the global flavor-defining charge-labels T'_3, B', etc., should be additively-conserved in most electroweak processes. This is just another way of saying that e, μ and τ type lepton-flavors are additively-conserved in most electroweak processes.

40. In general, weak-interaction flavor, or mass, eigenstates of fundamental fermions (quarks or leptons) do not exhibit significant mixing. In particular, lepton flavors, with the probable exception of neutrinos, do not seem to mix. Similarly, the weak-interaction quark eigenstates given by d', s' and b', where

$$\begin{pmatrix} d' \\ s' \\ b' \end{pmatrix} = \mathbf{V} \begin{pmatrix} d \\ s \\ b \end{pmatrix},$$

and **V** is the unitary Kobashi-Maskawa (KM) matrix, do not mix. But, the so-called strong-interaction flavor, or mass, eigenstates d, s and b obviously do mix when weak interactions are taken into account.

41. R. D. Peccei, "Why There Should not be a Fourth Family", in *The Fourth Family of Quarks and Leptons, Second International Symposium*, (Annals of the New York Academy of Sciences, New York, Vol. 578, Edited by D. B. Cline and A. Soni, 1989), p. 264.

42. L. D. Landau and E. M. Lifshitz, *Electrodynamics of Continuous Media*, Vol. 8 of *Course of Theor. Phys.*, (Pergamon Press, Oxford, 1975), pp. 58–59.

Index

baryon number B, 13, 15, 18, 22–27, 53–55, 67–73, 92
   as a function of the dimensionless parameter v, 19, 52
   as a linear function of Q, 24, 52
   as a quadratic function of Q, 24
   as a two-scalar, 25–27
   as part of the global hypercharge, 15, 32, 48
basis of family membership, 62–63
beauty, 46, 47–48, 60, 67–73, 93

Cartan subalgebra, 13–15
Cayley-Hamilton theorem, 3, 5, 7, 74
charm, 46, 47–48, 60, 67–73, 94
conservation of fermions, 2
constants of the motion, 2, 18, 32, 72–73

dependent flavor-defining charges, 18
distinguishing quarks and leptons, 6, 7, 9, 11, 12, 19, 30, 31, *See also* leptons as "inverted" quarks
distinguishing up- and down-type flavors, 39–41, 67–73, *See also* "up"-"down"-type flavor dichotomy

effective "quantization," 19, 33, 39–48, *See also* electric charge quantization
   and the components of the vector $\mathbf{Q}$, 50–56
   and the physically acceptable $\mathbf{U}$-vectors, 39–47
   of the parameters v and v', 52–55
electric-charge quantization, 24, 37, 40, 52–56, *See also* strong-color multiplicity
entropy, 79
even-dimensional representations of $\mathbf{F}$, 91
ex post facto "predictions," 7, 13, 19, 45, 67–73, 74–76, 85–87

families defined, 20, 21
   and composite models, 1
   and family gauge-symmetries, 1
   and flavor-doublet lowering matrices $\mathbf{R}^{-1}$, 62
   and flavor-doublet raising matrices $\mathbf{R}$, 61
   and random dynamics at the Planck level, 1, 77
   and superstrings, 1, 76
family hierarchy, 20–21, 59–63
family kinship, 48, 57–58, 63

in $SU(5)$, 1, 80–82
in the standard model, 1, 88, 89
family problem, 1, 49, 76–77, 81
fermion number, 1–12
    as a 2-scalar, 27
    as eigenvalues of **F**, 4–7, 11, 12
    basis for introducing, 3
finite two-element group, 5, 8, 56, 57, 66–69, 83, 87
first family, 20–30, 42–44, 62
flavor-asymmetric matter, 3, 67, 77–78
flavor-defining quantum numbers, 3, 7, 11, 13, 15–19, 31–33, 39–49, 59–61, 67–73, 80–82, 85–89, *See also* 2-scalars
flavor-doublets, 6–12, 16–19, 20–33, 39–48, 56–63, 66–70, *See also* "up"-"down"-type flavor dichotomy
flavor eigenstates, 15, 17, 32, 66–70
flavor-symmetric matter, 3, 67, 77–78
form of the vectors **U** and **V**, 39, 40
four flavors per family and four charge-measurement or flavor directions, 21–23, 34, 36

Gell-Mann–Nishijima formula, 15, 32, 40, 41, 71, 72, 92
general symmetry-evolution principle, applied to charge-vector triads, 56–58
"generalized fermion-number" **F**, 1, 4–10, 11, 12, *See also* new organizing principle for fundamental fermions
    as a function of the parameter v or v', 7–9, 50
    as a square-root of unity, 5, 74
    characteristic equations of, 4, 6, 7, 11, 12, 27, 28, 51
    eigenvalues of, 3–7, 27, 28, 51, 68, 69
    eigenvectors of, 4, 6, 27, 28, 51, 68, 69
    new degrees of freedom associated with, 9, 12, 29, *See also* geometric interpretation of three quantum dichotomies
    symmetries $S(I)$ or $S(II)$ associated with, 56, 57, 60–62, 66, 68, 69, 85–87
    trace and determinant of, 5
geometric interpretation of three quantum dichotomies, 12, 29, 30, 80–82
    matter-antimatter dichotomy, 2, 3–7, 9, 11, 12, 28, 29, 68, 69, 80, 82–84, 86
    quark-lepton dichotomy, 6, 7, 11, 12, 30, 68, 69, 80–82
    "up"-"down"-type flavor dichotomy, 3, 6, 7, 9, 11, 12, 29, 68, 69, 80–82
geometry and topology of the 2-space, 21, 22, 34–38
global charge-relations of quarks and leptons, 72
global conservation-laws of quarks and leptons, 72–73

INDEX

"good" quantum numbers, 2, 3, 7, 11, 15, 16, 18, 32, 67–73

Hilbert space, 2, 6, 10, 14, 79
hypercharge, 15, 18, 32, 40, 41, 48, 67–73, 85–87, 88–89, 92

interpretation of scalar components of 2-vectors, 10–86, See also 2-scalars
- **F**, 56
- **F**$_{\text{diag}}$, 10
- **g**, 10
- **R**, 63–65
- σ_z, 10
- **U**, 16–18, 39–41, 47, 59–61, 67, 68, 69, 85–87
- **V**, 16–18, 39–41, 68, 69, 85–87

isolated physical systems, 2, 15, 18, 28, 57, 58

labeling conventions, 21–23, 50–51
lepton number, 22–27, 53, 54, 67–73, See also baryon number
- as a function of the parameter v', 19, 52
- as a linear function of Q, 24, 52
- as a quadratic function of Q, 24
- as a two-scalar, 25–27
- as part of a global hypercharge, 32, 48

leptons as "inverted" quarks, 30, 31, 36, 38, 40, 42, 45, 47, 52, 55, 59–61, 66, 68–70, 72

linear independence of **U** and **V**, 12

Lorentz 4-scalars, 2, 7, 11, 16, 27, 71

Lorentz group, 10

matter-antimatter dichotomy, 2–12, 27–30, 34, 36, 38, 47, 60, 66, 68, 69, 79, 81, 83, 84, See also geometric interpretation of three quantum dichotomies
- and charge conjugation **C**, 6–9, 11, 28–30, 83, 84
- and $-\sigma_x$, 8
- and the necessity for a non-Euclidean metric, 10, 11
- effect on flavor-doublet lowering-matrices, 62
- effect on flavor-doublet raising-matrices, 61
- effect on 2×2 matrices, 8, 63–65, 83, 84
- effect on 2-scalars, 25–27, 68–71, 83, 84
- effect on vectors representing flavor-doublets, 8, 45, 47, 51, 60, 61, 68, 69, 83, 84, 86
- eigenvalues of $-\sigma_x$, 43, 86
- eigenvectors of $-\sigma_x$, 43, 86
- geometric interpretation of $-\sigma_x$, 29, 30, 34–38
- in $SU(5)$, 81

minimal set of independent flavor-defining charges, 17, 18, 39, 40

Möbius topology and $S(II)$-type vector-triads, 57

new organizing principle for fundamental fermions, 1, 6, 7, 11–12, 74

"normalized" baryon or lepton numbers, 27
number of families N, 1, 20, 21
  and composite models, 1
  and helium nucleosynthesis, 1
  and random dynamics at the Planck level, 1, 77
  and $SU(5)$, 1, 81
  and superstrings, 1, 76
  and the new geometry, 2, 7, 12, 13, 19, 33, 42–45, 59–63, 68–69
  and Z^0 decay width, 1

orthogonality of matter and anti-matter flavor-doublets, 28, 29

physical interpretation of the 2-space, 13–19
Planck-level dynamics, 1, 13, 76–78
probability amplitude(s), 2, 6
probability density, 10

quark-lepton dichotomy, 12, 19, 30, 36, 38, 47, 68, 69, 79, *See also* geometric interpretation of three quantum dichotomies
  in $SU(5)$, 82
quarks as "inverted" leptons, 30, 31, 36, 38, 40, 42, 45, 47, 52, 55, 59–61, 66, 68–70, 72

representing individual flavors and flavor-doublets by triads of 2-vectors, 12, 16, 17

second family, 21, 44–48, 59–63

second-law of thermodynamics, 57
selection-rules, 48, 49, 71, 94
strangeness, 15, 18, 32, 33, 46–48, 67–73, 92
strong-color multiplicity, 19, 54–56, 64
superstrings, 1, 76
symmetry-evolution of vector-triads, 49, 57, 58, 78

third family, 21, 44–49, 59–63
third-component of global isospin, 14, 15, 18, 40–42, 47, 60, 67–72, 85–87, 92
transformations of flavor doublets, 66
truth, 46–48, 67–73, 93
2-scalars, 4, 7, 10, 11, 14–18
  and scalar-products of 2-vectors, 10, 11, 14–18, 23–29, 31, 32, 41–49, 53–55, 60, 61, 68, 69, 71–73
  components of **F**, 7, 9, 10, 50, 56, 91
  components of **Q**, 4, 7, 8, 9, 11, 14, 15, 16–19, 38, 50–55, 68, 69, 71, *See also* electric charge quantization
  components of **U**, 15, 17, 18, 39–45, 59–61, 67–69, 71, *See also* charm and beauty
  components of **V**, 15, 17–18, 68, 69, 71, 87
Q^2-charge, 10, 11, 14–18, 23, 25–29, 31–33, 53–55, 71, 68, 69, *See also* baryon number; lepton number
u-charge, 17–18, 39–45, 47, 59–

INDEX

61, 67–79, 87, *See also* third-component of global isospin

U^2-charge, 15–18, 31, 32, 41–48, 60, 61, 67–73, *See also* strangeness; truth

U•V-charge, 15–18, 31, 32, 41, 68–69, 71, *See also* hypercharge

V^2-charge, 15–18, 31, 32, 41, 67–69, 71

2-space, 4, 6, 7, 10
- and associated one-dimensional invariant vector-spaces in, 21–23, 35, 38, 42, 43, 60, 68, 69, 75, 87, 93 *See also* X_1 and X_2 coordinates
- and pseudorotations $SO(1,1)$, 10
- flavor directions in, 23, 34–36, 75
- graphical representation of, 34–38
- non-Euclidean metric **g**, 10
- on a Euclidean plane (E2), 21–23, 34–38, 43, 47, 66
- physical and unphysical quadrants, 35–38
- topologically-distinct regions in, 37–38, *See also* physical and unphysical quadrants
- X_1 and X_2 coordinates, 21–23, 29–31, 34–36, 42, 43, 45

2-vectors, 4, 6, 7, 11–13
- $\mathbf{Q}_l$, 11, 12, 15, 17, 23, 26, 27, 31, 33, 37, 38, 48, 50, 66, 69, 85
- $\mathbf{Q}_q$, 11, 12, 15, 17, 23, 26, 31, 33, 37, 38, 48, 50, 66, 68, 83, 85–87

U, 12–19, 31–33, 39–48, 59–61, 66, 68–69, 85–87

V, 12–19, 31–33, 39–48, 59–61, 66–69, 85–87

"up"-"down"-type flavor dichotomy, 6, 9, 11, 12, 17, 20, 21, 29, 79, *See also* geometric interpretation of three quantum dichotomies

in $SU(5)$, 81

weak-hypercharge and weak-isospin, 24, 25, 80–82, 88, 89

About the Author

Gerald Fitzpatrick has a background in geophysics (Colorado School of Mines) and physics (University of Denver). He is a member of the American Association for the Advancement of Science, the American Physical Society (APS)-Division of Particles and Fields (DPF), the American Society for Nondestructive Testing, the International Society for Optical Engineering, and the New York Academy of Sciences. On the vocational side of his career, for the last 30 years he has specialized in applied physics doing government- and industry-sponsored research in linear and nonlinear physical-acoustics, optical- and acoustic-holography, laser-vibrometry and magneto-optic/eddy-current imaging. He has published over 35 research papers and holds a number of U.S. and foreign patents. For his invention and development of magneto-optic/eddy-current imaging technology he, and his current employer Physical Research Inc., were honored by a 1990 R&D 100-Award (by R&D Magazine), a 1991 Photonics Spectra Circle of Excellence Award (by Photonics Spectra magazine) and the 1994 Technology 2004 Award of Excellence in Technology Transfer (sponsored by NASA, NASA Tech Briefs and the Technology Utilization Foundation in cooperation with the Federal Laboratory Consortium for Technology Transfer). On the avocational side of his career, and for the same 30 year period, he has immersed himself in the study of elementary-particle physics in an attempt to address the rhetorical question posed by I. I. Rabi in 1947 when confronted with the discovery of the electroweakly-interacting spin $\frac{1}{2}$ muon: 'Who ordered *that*?' Today the scope of this question has been greatly expanded. It is now referred to as the "Family Problem" since the muon is only one member of the second family, of three known families, of "fundamental" fermions (quarks and leptons). From 1975 to the present time he has been presenting his evolving ideas on this subject at APS general meetings and meetings of the DPF. The subject matter of this book is a summation of the author's long-term research effort and represents his best "answer" to Rabi's question.